P9-DZA-718

THE FORMAL ORGANIZATION

ISSUES AND TRENDS IN SOCIOLOGY

a series of
The American Sociological Association

INTERGROUP RELATIONS
Edited by Pierre van den Berghe

THE FORMAL ORGANIZATION
Edited by Richard H. Hall

The Formal Organization

EDITED BY

Richard H. Hall

BASIC BOOKS, INC., PUBLISHERS

NEW YORK LONDON

LIBRARY OF CONGRESS CATALOG
CARD NUMBER: 72-76921

cloth SBN 465-02491-2
paper SBN 465-02492-0

Manufactured in the United States of America

72 73 74 75 10 9 8 7 6 5 4 3 2 1

The Authors

MICHAEL AIKEN is Professor of Sociology at the University of Wisconsin and co-author, with Jerald Hage, of *Social Change in Complex Organizations*. He is currently studying power relationships in communities.

PETER M. BLAU is Professor of Sociology, Columbia University and President-elect of the American Sociological Association. Among his books are *The Dynamics of Bureaucracy*, *Bureaucracy in Modern Society*, and *Exchange and Power in Social Life*. He is also the co-author of *The American Occupational Structure* and *The Structure of Organizations*.

GEORGE BRAGER is Professor of Social Work at Columbia University. He was responsible for planning and implementing the program of Mobilization for Youth. He is co-author of the forthcoming book, *The Process of Community Organizing*.

MOHAMED EL-ASSAL is Professor of Sociology at California State University at San Diego and co-author of a forthcoming book, *Focal Concerns in Mass Communication: Adapted Reading*.

JERALD HAGE is Professor of Sociology at the University of Wisconsin. With Michael Aiken, he is co-author of *Social Change in Complex Organizations*. He is co-author of two other volumes, *Techniques and Problems of Theory Construction* and *Organizational Systems: A Text-Reader in the Sociology of Organizations*, currently in press.

RICHARD H. HALL is Professor of Sociology at the University of Minnesota. His works include *Occupations and the Social Structure* and *Organizations: Structure and Process*.

T. ROBERT HARRIS is a Research Associate in Sociology, School of Urban and Public Affairs, Carnegie-Mellon University. He is primarily interested in political sociology.

D. J. HICKSON is Ralph Yablon Professor of Behavioral Studies, Organizational Analysis Research Unit, University of Bradford Management Center, England. He is co-author of *Writers on Organizations* and *Power in Organizations* (forthcoming).

C. R. HININGS is Senior Lecturer in Sociology, Department of Industrial Administration, University of Aston, England. He is co-author of *Writers on Organizations* and *Power in Organizations* (forthcoming).

MARK LEFTON is Professor and Chairman, Department of Sociology, Case Western Reserve University and co-author of *Schizophrenics in the Community*, *Women after Treatment*, and *Hospitals and Patients*: *A Theory of Clients and Organizations*.

GEORGE A. MILLER is Assistant Professor of Sociology, University of California, Los Angeles. He is co-author of *The Sociology of Organizations*: *Basic Studies*.

D. S. PUGH is Professor and Chairman of Organizational Behavior at the London Graduate School of Business Studies. Among the books he has co-authored are *Exercises in Business Decisions*, *Writers on Organizations*, and *Organization Theory—Selected Readings*.

WILLIAM R. ROSENGREN is Professor and Chairman, Department of Sociology, University of Rhode Island. He is co-author of *Hospitals and Patients* and *Organizations and Clients*: *Essays in the Sociology of Service*.

JOSEPH W. SCOTT is Associate Professor of Sociology and Director of Black Studies, Notre Dame University. He is continuing his research on university structures and student protests and the sociological bases of "black power."

ARTHUR L. STINCHCOMBE is Professor of Sociology, University of California, Berkeley. His books include *Constructing Social Theories* and *Rebellion in a High School*.

C. TURNER is Lecturer in Sociology, School of Social Studies, University of East Anglia, England.

STANLEY H. UDY, JR. is Professor of Sociology, Dartmouth College and the author of *Organization of Work* and *Work in Traditional and Modern Society*.

Preface

This volume is one of a series on Issues and Trends in Sociology sponsored by the American Sociological Association. Each is the product of a distinguished editor's work. His task has been to assemble from past numbers of the Association's periodical publications the accumulated thought on a selected topic, supplementing those contributions with materials from other sources as needed, and to examine and interpret the state of knowledge represented in the collected papers with reference to its implications for current intellectual and social trends. By this means the Association has sought to foster both the advance of scholarship and the understanding of an important issue of the day. We take pleasure in presenting this volume on behalf of the American Sociological Association.

Contents

PART III

The Impact of Internal Factors on Organizations

PART IV

Some Consequences of Organizational Structure

Figures and Tables

TABLE

TABLE

TABLE

THE FORMAL ORGANIZATION

Introduction

Most of the work on this book was done in Washington, D.C. It is an exciting, yet frightening place. The excitement comes from seeing the government, lobbyists, protest groups, and so on in action. The fright comes from seeing these same organizations in inaction. Excitement also comes from seeing an organization change, while fright ensues from wondering what the effects of the change will be. The direction of these emotions is dependent on one's personal perspective. The important thing is that most of what happens or does not happen in society has an organizational basis. The organization is the instrument of change and the great resistor of change.

As subject matter, then, organizations are central to understanding and analyzing society. The sociological study of organizations is beginning to reflect this. The issues and trends in the field now concern such things as the impact of organizations on their members and the wider society, interactions between organizations and effects of such interactions on the organizations and the society, and the conditions in the environment which impinge upon the organization. The sociology of organizations is concerned with these broader issues. It is also vitally concerned with understanding organizations *qua* organizations. The subject matter itself is a social phenomenon of intrinsic interest.

The purpose of this volume is to present what is essentially a "status report" on the state of this particular subfield within sociology. Any such status report is by definition out of date as soon as the ideas are transmitted to the typewriter, because other ideas are being similarly transmitted and cannot thus be included at a particular point in time. Nevertheless, there is a need to assess where the field of organizational analysis now is and where it is likely to go. In order to do this, the articles selected for inclusion in this volume will be recent and for the most part based on comparative empirical research. The criterion of recentness is used with only minimal hesitation. While certainly it is proper to pay homage to and give thanks for seminal works which have appeared at an earlier historical period and while many provocative ideas have not seen the light of an honest research day, recent research has the distinct advantage of indicating what the directions of research have been. They also are probably excellent indicators of where it is likely to go in the next few years, since like the subject matter itself, the structure of this subfield will in many ways help to determine what happens to it over time. At the same time, of course, the references in the articles themselves constitute valuable bibliographical sources.

The criterion of comparative empirical research is employed for, perhaps, obvious reasons. The field has moved beyond the case study. This has happened in most of the subfields of sociology. The research techniques employed in the papers that follow indicate the rich diversity of approaches that can be brought to bear in organizational analysis. They also indicate that closure has not been reached on the manner in which many key variables should be measured. Indeed, there are some indications that different measures of the same phenomena yield somewhat inconsistent results. While these inconsistencies are healthful in the sense of leading to continued growth, they are frustrating when it comes to interpreting results. Despite this problem, there is a growing amount of agreement on certain key relationships, as will be indicated in the sections which follow.

This set of readings will be organized around three basic questions about organizations—What are they? How are they? and So what?

The first question will be answered by examining the characteristics that are typically included in analyses of organizations. These characteristics are assumed to be present in varying degrees in all organizations. This assumption itself requires additional comment since it contains a major assumption which will underlie all of the discussion which follows. The assumption is that all organizations share characteristics which allow them to be analyzed as a distinct class of phenomena. This means that certain crucial characteristics of the various anti-war organizations can be compared and contrasted with similar characteristics of the United States Army. The presence of these common characteristics means that the distinctions between voluntary and non-voluntary organizations and between profit and non-profit organizations are meaningless unless the distinction makes a difference insofar as organizational characteristics are concerned. It may be, for example, that the personnel procedures of a local service station are much less formalized than those of an exclusive private club in its membership selection. Insofar as organizational characteristics are concerned on this variable, the club would be more like other organizations with similarly stringent personnel policies than other voluntary organizations which are not so rigid in their entrance requirements.

If a discriminating and accepted typology of organizations were available this issue would not be as bothersome as it is. Unfortunately, such a typology is not available so that one essentially has to classify organizations as he goes along. Depending on the characteristic(s) under consideration, therefore, every organization has more or less of this variable present and can therefore be classified along with all others. It is a hope that future research will yield sufficient information regarding the relationships among organizational characteristics so that a truly empirically based typology can be developed. Until that time it is probably wise not to rely on *a priori* typologies which may be more misleading than useful and to rely upon variations in structure and process as means of classifying organizations. There is not total agreement about exactly which variables should be included in a comprehensive analysis. There is enough agreement, however, so that meaningful comparisons between organizations can be made and the nature of organizations explicated.

After examining the nature of organizations, the analysis will turn to the issue of how organizations come to be as they are. Most of the discussion will be based around organizational structure, since this is the framework in which action occurs. The issue will be handled in two sections, since the evidence is increasingly suggesting that there are two major classes of factors which contribute to the form which organizations take. *External* and *internal* considerations affect the organization. The nature of these factors is such that a static view of organizations is impossible, except under the unlikely conditions of no change in the total social system.

This distinction between internal and external matters should be further developed and extended for the sake of clarity and the general theoretical perspective which is the basis for this volume. The perspective which will be taken in organizing these readings is that organizations can most fruitfully be understood through viewing them as goal-oriented social systems. This is in keeping with contemporary thought on organizations as expressed in the writings of James Thompson and Daniel Katz and Robert Kahn.[1] This approach avoids the pitfalls of either viewing the organization as a totally closed system with all behaviors being based around pursuit of a commonly agreed upon goal or as a totally open system in which the organization simply responds to external threats and internal conflicts with no direction at all. The current perspective is that organizations are formed in the first place to accomplish something and that the members of the organization are in it due to a variety of ties to this goal or set of goals.[2] These goals become constraints on the decision making process within the organization. At the same time they are affected by a series of exigencies which may or may not have been anticipated at the time of the organization's formulation. These exigencies force the organization to deflect some of its resources in an attempt to cope with the new situations. While this occurs operative goals emerge as guidelines for action. These operative goals essentially are the standards by which decisions are made for further allocation of resources. They can conceivably be quite distinct from the originally developed official goals.

The kinds of exigencies faced come from internal and external pressures. Internal pressures, such as the development of new power bases within the organization, new militancy among segments of the organization, or conflicts which arise which require resolution, force the organization to pay some attention to these internal problems and thus commit some resources to these issues. The very fact of doing this creates new operative goals, since these internal conditions become criteria for decision making. The same general point is true for the external side of the equation. As new governmental regulations, changed demographic conditions, new competitors, advances in technology, and a new breed of student appear on the scene the organization must react if it is to stay a viable entity. This reaction again is a source of resource deflection and new operative goals. While these factors have an important impact on the organization, except under cases of extreme emergency, the basic goals and their derived operative guidelines remain as bases for decision making and structuring in the organization. Extreme cases would either result in the total redefinition of what the organization is supposed to do or, in a somewhat unlikely situation, the dissolution of the whole operation.[3]

The distinction between external and internal factors is essentially a heuristic one. Most of the kinds of things which are taken as internal pressures on the organization have their sources outside the organization. For example, the role of the professional in the organization will be examined and it will be shown that the presence of this type of personnel has an impact on the structure of the organization. Professionals and their beliefs and behaviors obviously do not just spontaneously emerge in the organization, but rather are brought to the organization from outside. The same thing is true when other members of organizations become militant or passive about the state of their existence in that the ideas usually have an origin outside the organization. While internally based factors may affect changed orientations toward the organization, even in the most secure prisons or asylums there is some kind of input from the outside which interacts with the felt dissatisfactions inside and leads to a new perspective on the situation and possibly action to change the situation. It is conceivable, of course, that organizations would be insulated from external pressures and from changes in their internal system so that these factors would not be a consideration at all. In these cases, of course, there would probably be little, if any, deflection from the official goal.

This brief excursion into these theoretical matters was intended to set the basic framework for the analysis which follows. It is important to note that this is not verified theory, but rather reflects the direction that recent evidence about organizations suggests. The recent evidence comes largely from analyses of organizational structure.

Up to this point, the discussion has been based around change and differences between organizations. While change and differences are an important component of the analysis and of the nature of organizations, it is evident that important regularities are also present. The literature to be discussed in the balance of this volume will suggest that under common external and internal conditions, organizations will assume the same structural forms. Since the external and internal conditions themselves vary, organizational structures will similarly vary. The "state of the art" is such that this variation cannot be expressed in precise mathematical terms, but the data and verbal arguments presented will make a strong case for systematic and predictable variations among organizations.

The final question, "So what?" is by far the most important one. While analysis of what organizations are and how and why they vary is an interesting intellectual exercise for those who enjoy that sort of thing, there is no automatic pay-off in terms of relating organizations to the wider society or to their members. That is a separate task. The purpose of this section is to indicate some of the ways in which the form of an organization makes a difference in what it does to and for people, the organizations themselves, and the larger social structure.

In keeping with the orientation expressed at the outset of this discussion, these readings will be based around recent research. This puts some severe constraints on the material which can be included. Since the research process is relatively slow itself, and the reporting of research findings a relatively unpredictable phenomenon, some issues which have captured the concern of many segments of the population will not be represented. For example, the role of organizations in the development of environmental quality or disintegration is certainly being examined, but, as yet, there are no research reports available which provide any really systematic evidence. The same is true in regard to organizations designed to oppress or benefit various segments of the population. In these areas it is tempting to take whatever is available to indicate relevance. The decision here is not to go that route, since the writings which are available are overwhelmingly nonobjective and based on extremely limited evidence. What will be attempted here is to indicate the implications from contemporary evidence for some of these major societal issues. Since organizations are central to developing, developed, and over-developed societies, their impact on these societies must be systematically and continually examined. The same is true of their impact on their own members, since particularly in the latter two kinds of societies such a major part of a person's life is spent in organizations.

Specifically in this final section the impact of varying organizational structures for three parts of the social system will be examined.

The first aspect of the social structure which will be examined is the organization itself. Although at first glance it seems absurd to look at the way organizational structures affect the organization itself, a more careful analysis of the idea reveals its centrality to the whole study of organizations. Organizational structures are created by the members of an organization. Once created such structures serve as the basis for future operations until the structure is changed. Each change serves as a further guideline for the organization and its members. The evidence to be discussed will show that if an organization is structured in a particular way, certain conditions, problems, and dilemmas must be faced which are different than they would be if the organization had an alternative structure.

The second component is the members of the organizations themselves. It will be shown that different structural arrangements affect different kinds of people in different ways. While the research evidence here is not as complete as would be desirable, that which is available suggests that organizational structural arrangements interact with individual characteristics leading to behavioral and attitudinal differences on the part of the individuals depending on the mix of the factors involved. In simple terms, there can be too much, too little, of even the right amount of organizational structure for the individual. This appears to be true from the perspective of the individual's satisfaction and happiness in his work and in terms of the organization's perspective of optimizing the contributions of each member.

The final relationship to be discussed is that between organizations and the wider society itself. It will be suggested that differing forms of organizations have differing impacts on the society of which they are a part. It is in this area that the evidence is the weakest and the research most difficult. The difficulties here are quite understandable. If the assumption is made that social change at either the macro or micro level has a multicausal basis, tracing the impact of organizations as opposed to other sources of change and stability is a difficult task. The task is made even more difficult by the fact that good indicators of social change itself are not readily available or in wide use. The articles in this section will suggest that the presence of alternative organizational forms makes a difference in terms of how organizations respond to their environment and how they participate in the society as agents of or reactors against social change.

Given the organizational density in contemporary Western society and the evident imperfections in that society it is this area which should serve as a major focus for research by students of organizations. In a very real way organizations have become the heart of our current way of life. Much of what is wrong with the society has its origins in the organizations which comprise it. Much of what is good has similar origins. However that may be, the organizational responses to problems of quality of the environment, poverty and inequality, individual expression, and many other contemporary issues must be identified and analyzed. Such matters, though as of now they are but weakly conceptualized, represent some of the important areas for future research. The field of organizational analysis has had a rather late development, as is suggested by the recent dates on most of the papers included here.

In addition to the kinds of omissions alluded to above, the reader will notice that there are several other areas not included in this volume. Since the readings are designed to deal with organizational structures, relatively little mention will be made of organizational processes. While the distinction between structure and process is only useful from an analytical perspective, since structure itself is in a constant state of change, the analysis of processes such as conflict and cooperation, decision making, and supervision have unfortunately generally proceeded without benefit of considering the structural underpinnings which serve as their basis. Structural analysis has similarly tended to ignore processes. The consequences of this mutual lack of attention has led to a condition wherein both kinds of literature seem strangely unconnected. While there has been too little connection between structural and processual analyses, it should be pointed out that the entire area of social processes, while widely discussed, has received little systematic attention in the research process, probably because it is so difficult to conceptualize and operationalize. Although everyone has experienced conflict in an organization or between organizations, when it comes to measuring its scope and intensity there are few usable indicators for serious research.

Minimal attention will be paid to the individual in the organization. His behavior, currently labeled organizational behavior, has been the subject of intensive investigations from the perspective of differing motivational properties, reactions to and participation in alternative supervisory styles, and behavioral change through sensitivity and other such individual behavior modifying experiences. While there is no intent to diminish the importance or soundness of much of the research in this area, this particular book is designed to deal with organizational structures. This intent is coupled with my own belief that structure is actually more important to the understanding of what goes on in an organization than the particular individuals involved. While the activities of individuals are certainly an important input to organizations, the very nature of organizations is such that they are intended to minimize the influence of individual variations (this is not meant to imply that organizations are therefore also designed to dehumanize the individual—in fact the opposite could be the case). Structural characteristics are an important consideration in understanding how an individual does react and behave in an organization. Similarly, the kinds of members in an organization make an important difference in how it is structured. These kinds of relationships will be explored, with the stress on the organizational structure, rather than the individual himself.

The final intentional omission is a vital one. All organizations are concerned with how well they are doing their job, or their effectiveness. The growing amount of attention being paid to this area indicates its importance in organizational analysis. This attention also indicates the absence of clearly identified models by which effectiveness can be judged. Outside of profit-oriented organizations, and even here there are good indications that much uncertainty exists about what effectiveness really means, organizations themselves have difficulty in assessing how well they are performing. That this is crucial for organizations can be ascertained by the fact that organizations are designed to accomplish something. That they have difficulty in determining whether they do or do not accomplish is clear evidence of the complexity of the effectiveness issue. Some of the articles will discuss effectiveness, while others will not. The basic position taken here is that there are multiple criteria of effectiveness and multiple means by which organizations can be effective. There is no one best way for all organizations to structure themselves. More importantly, within the same general sphere, such as industry, the same point appears to hold. Thus, while effectiveness is really a common concern of all students of organizations, it will be an implicit concern here.

The next section will deal with the nature of organizations. It will indicate the components which are usually included in analyses of organizational structure and its antecedents and consequences and will serve as the common "ground" for the discussions which follow.[4]

NOTES

1. James Thompson, *Organizations in Action*, New York: McGraw-Hill, 1967; and Daniel Katz and Robert L. Kahn, *The Social Psychology of Organizations*, New York: John Wiley & Sons, 1966.
2. For an extended discussion of this see Amitai Etzioni, *A Comparative Analysis of Complex Organizations*, Glencoe: Free Press, 1961.
3. See Daniel Sills, *The Volunteers*, Glencoe: Free Press, 1957.
4. The overall perspective taken here follows Richard H. Hall, *Organizations: Structure and Process*, Englewood Cliffs, N.J.: Prentice-Hall, 1972.

PART I

The Nature of Organizations

This will be a brief section. Its brevity is not due to the fact that all of the intricacies of organizations have been resolved, nor that there is agreement on the most crucial components. Rather, the two papers included are indicative of the dominant approaches to organizational structures at the present time. This dominance is both conceptual and methodological, even though there is enough divergence, conceptually and methodologically, to indicate plenty of room for future development.

The commonalities of the papers are clear. They pay the proper homage to Weber as the sociological progenitor of organizational analysis and both then procced to point out some of the insufficiencies in the Weberian model.[1] They also approach the components of structure as variables which do not necessarily vary together, even though typical patterns of relationships among variables may be found under common conditions. Another commonality, and this is an unfortunate one, is that the organizations investigated in no way represent a sample of the organizational universe. Udy's data come from non-industrial societies. The Pugh, *et al.* research was carried out in the English Midlands, where the researchers themselves were located. Saying this is not intended to diminish the contributions of these projects or any others with the normal and probably unavoidable time and geographical limitations. Rather it is intended to indicate one of the major difficulties in conducting organizational research, since there is no established universe from which a sample may be drawn or a selection of organizations compared. An important common limitation of the papers is that no non-work organizations are included.

We therefore do not know whether the dimensions of organizational structure discussed would be at all relevant in voluntary organizations. It may well be, however, that a local chapter of a weight watcher's organization would score low on most of the dimensions discussed, with some voluntary organizations being more bureaucratized or having their authority more concentrated than many work organizations. This area is one in which there is a clear need for additional research.

The papers are attempts to utilize the variables central to organizational structure as they have been identified in the literature. Although the terminology in the papers is different, the phenomena it refers to are quite similar. For example, the Pugh, *et al.* work refers to centralization, standardization, and formalization. These are extremely close to Udy's hierarchical authority structure and specialized administrative staff. Each contains elements not included in the other, but both papers together indicate the range of variables which have been of most central concern to organizational analysis.

A major difference between these papers is in the methodology utilized. Udy uses a simple presence/absence measure. Pugh, *et al.* use descriptions of the organizations supplied by knowledgeable people within the organization.[2] While it would be nice to have data indicating the manner in which these approaches yield similar or different results, such data are not available. There are some indications, however, that the approaches do produce somewhat different results in that an organization which scores "high" on a variable in the Pugh technique may not show up as "high" when the Hall and Aiken-Hage approach is used.[3] These difficulties have not been systematically examined, however, and the cause of the possible differences must remain in the realm of speculation.

An important point in the consideration of these methodological differences is the fact that the approaches are actually measuring different things, even though they are treating common variables. It has been long established that the behavior of members of organizations does not conform to the letter of the organizational law. The Pugh approach is oriented toward the nature of the law. The Hall and Aiken-Hage approach gets at perceptions of how the organization is structured. That these are different is really a foregone conclusion. Since there does seem to be a good case for the validity of both measures, the differences which may obtain from the different methodologies are apparently based on real differences in what is measured. These considerations, of course, raise the interesting questions of how much, under what conditions, and in what pattern deviance from official standards occurs in organizations. This, of course, is a whole separate issue.

These methodological considerations are important for the rest of this volume. The state of the field is such that these are the primary methods by which information about organizational structure is obtained in the field as will be evident in the readings which follow. Direct observation and recording of behavior has thus far proven to be prohibitively expensive for comparative research and other methods just simply have not been developed to the point of wide usage. Part of the methodological problem is minimized when the fact that the research findings themselves are quite consistent is considered. This fact suggests that despite the differences in measurement, consistently patterned relationships are being found.

Udy's work is really the first to take a careful and comparative empirical look at the manner in which the structure components of organizations hang together. His findings suggest an inherent instability in organizational structures.

The Pugh, *et al.* paper takes a truly empirical approach to determining the basic structural elements of organizations. After identifying what they call "primary dimensions of structure" from the literature and developing several measures of each, organizational scores on the multiple measures were then factor-analyzed. The results of this analysis revealed four underlying dimensions of structure. These are: The structuring of activities, concentration of authority, line control of workflow, and size of supportive component. The authors suggest that these underlying dimensions are characteristic of all work organizations. They also suggest that this throws the traditional bureaucratic model into disutility. Organizations are not bureaucratic or nonbureaucratic. Instead, they vary along the several dimensions specified. Later sections of this book will try to identify the reasons for the variations identified.[4]

NOTES

1. References in the footnotes of both papers should provide ample references to Weber's works and those which have followed. Other papers in this volume contain additional references to the classic writers.

2. Another important methodology used in the comparison of organizations is to utilize the perceptions of members of the organization about the organization. This method can be seen in the papers of Hall, Hage and Aiken, and Aiken and Hage which appear in later sections.

3. Conversations with David Hickson have suggested that these differences do exist, based on the application of both methods to the same organizations.

4. An obvious, but often overlooked, point should be noted. Organizations vary internally in terms of how they are structured. Not all segments of an organization exhibit equal structuring of activities or concentration of authority. For the most part, however, we will treat organizations as units, recognizing that a more sophisticated analysis would include these internal variations.

1

STANLEY H. UDY, JR.

"BUREAUCRACY" AND "RATIONALITY" IN WEBER'S ORGANIZATION THEORY: AN EMPIRICAL STUDY

Max Weber's "split personality" as sociologist, on the one hand, and transcendental idealist historian, on the other, has from time to time occasioned comment in the literature.[1] This duality of posture in Weber's work appears in particular to have had some rather interesting consequences for the lines along which contemporary organization theory has developed. It is the purpose of this paper to discuss some of these consequences, together with certain problems to which they lead, and to explore a possible solution to these problems through a small empirical study.

The dual nature of Weber's approach is perhaps nowhere so apparent as in his use of ideal types in the analysis of administrative structure. From a sociological point of view, his specifications for the "rational bureaucracy" superficially resemble the categories of a model. Yet on closer scrutiny they prove to be alleged concrete attributes, rather than variables or categories in a classificatory scheme. And, indeed, from a transcendental idealist point of view, Weber himself treats ideal types as substantive conclusions rather than methodological tools. In this sense, his specifications for the "rational bureaucracy" represent not so much a system of analytical categories as they do an attempt to capture the "spirit" of contemporary administration.[2] Despite their partially meta-

Reprinted from Stanley H. Udy, Jr., " 'Bureaucracy' and 'Rationality' in Weber's Organization Theory: An Empirical Study," *American Sociological Review*, 24 (December, 1959), pp. 791–795.

physical character, however, Weberian ideal types have a definite empirical flavor. Consequently they have proved highly useful in empirical work, and of course have exercised a well-recognized influence on contemporary organization theory.

Certainly few will deny that such influence has been on the whole extremely fortunate. But at the same time it has not been without drawbacks. The legacy of Weber is probably in great part responsible for the extreme degree to which organization theory has had to face problems of flexibility. For however ingenious and perceptive they may be, ideal types cannot be applied directly to the analysis of empirical data. Instead, the investigator must either recast the ideal type as a model by reformulating its specifications as a system of interrelated variables, or rest content to study the respects in which actual cases do not conform to it. The second alternative has been the one usually followed, and has lead to the development of an extensive body of theory dealing with "informal organization"—namely, departures from ideal-typical "rational bureaucratic" characteristics.

Such development itself, however, has produced a mounting body of evidence which suggests that some attention might fruitfully be given the first alternative—that of attempting to recast the original ideal type as a model. For it appears that many informal behavior patterns can at least partially be accounted for by problems which seem inherent in the formal (in this context, the ideal-typical) structure. Moore, for example, indicates that the formal structure itself provides sufficient conditions for the development of certain informal behavior patterns, pointing out that formal (in this context, the ideal-typical) structure. Moore, for example, arise which lead to a body of accumulated precedent, and that formal structure provides the framework within which most informal interaction takes place. Similarly, Blau stresses the development of informal patterns of behavior as a result of repeated adaptations of the organization to constantly changing conditions. From a somewhat different perspective, March and Simon indicate that this problem can be approached from the viewpoint of latent consequences of motivational devices, presenting an interesting comparison of the models which they find implicit in the work of Merton, Selznick, and Gouldner in this regard.[3]

Findings of this kind cast some doubt on the long run advisability of maintaining a rigid distinction between "formal" and "informal" administrative behavior; they suggest an altogether different approach, involving a general model which would provide an explicit context within which these various "adaptive" mechanisms could be presumed to operate. It was believed that a reformulation of Weber's original ideal-typical attributes as variables, together with an attempt to establish their actual empirical interrelationships, might provide the desired flexibility and thus prove to be a fruitful step toward the construction of such a model. Accordingly, seven of Weber's ideal-typical "rational bureaucratic" characteristics are reformulated below as rough "present *versus* absent" variables. Subsequently, the extent to which they are all likely to be present together in any formal organization is investigated through a comparative analysis of 150 organizations engaged in the production of material goods in 150 different nonindustrial societies. The societies were selected according to Murdock's criteria for his "World Ethnographic Sample." One organization was then drawn at random from those in each society on which data were available. The principal source of data was the Human Relations Area Files.[4]

Concepts, Hypotheses, and Results

A *formal organization* is defined as any social group engaged in pursuing explicit announced objectives through manifestly coordinated effort. Weber's "rational bureaucracy" is conceived as such an organization, which, however, possesses as well other characteristics. Seven of these characteristics have been selected for study, and reformulated operationally as follows:[5]

1. An *hierarchical authority structure* is considered to be present if the organization possesses three or more levels of authority.[6]
2. A *specialized administrative staff* is reported if any member of the organization is concerned solely with activities other than physical work.[7]
3. Rewards are considered to be *differentiated according to office* whenever this is reported as being the case, either with respect to the amount or kind of reward.
4. An organization is deemed to possess *limited objectives* if, in the case of the sample used here, it is exclusively concerned with the production of material goods.
5. A *performance emphasis* is considered to be present if the amount of reward to members is in any respect dependent on quantity or quality of work done.

6. *Segmental participation* is considered present if participation is based on any kind of mutual limited agreement.

7. *Compensatory rewards* are reported if members of higher authority distribute rewards to members of lower authority in return for participation.

The above reformulation, together with the idea that the variables involved do not always necessarily vary together in the same direction, was suggested by the work of Weber himself. An examination of Weber's writings reveals that organizations which he characterizes as "bureaucratic" are distinguished by an hierarchical authority structure, an administrative staff, and differential rewards according to office. These characteristics therefore are designated the *bureaucratic* elements of formal organization. Limited objectives, a performance emphasis, segmental participation, and compensatory rewards, however, do not occur in all of Weber's forms of "bureaucracy," but only in the "rational-legal" type. For this reason, these four characteristics are termed the *rational* elements of formal organization. One of Weber's main points, in effect, is that rational characteristics do not invariably tend to be associated with bureaucracy. On the contrary, he finds such a combination to be rare, except in contemporary Western society. Much of his work is devoted to exploring the historical conditions which made possible this peculiar combination of organizational features. A precise hypothesis of opposition between bureaucratic and rational elements in formal organization, however, is not clearly explicit in Weber's work. But in an examination of Weber's theory of the routinization of charisma, Constas finds such an hypothesis to be implicit, reaching the conclusion (allowing for some terminological differences) that established bureaucratic characteristics tend to be dysfunctional for the institutionalization of rationality.[8]

Thus the specific hypothesis may be proposed that, in formal organizations, mutual positive associations tend to exist between bureaucratic elements and also between rational elements, but that rational elements tend to be negatively associated with bureaucratic elements. Table 1–1 gives the values of Q (Yule's coefficient of association) resulting from an investigation of the mutual association of the seven relevant organizational characteristics in the sample of 150 formal organizations.

It is evident from inspection of the table that the results lend general support to the hypothesis. They are also consistent with the theory advanced by Constas. The seven characteristics sort themselves quite neatly into two sets, composed of bureaucratic and rational elements, respectively.

TABLE 1-1
Mutual Association of Seven "Rational Bureaucratic" Characteristics In 150 Formal Organizations

	A	B	C	D	E	F
B	+.79*					
C	+.79*	+.70*				
D	−.29	−.66*	−.16			
E	−.12	−.27	+.40	+.66*		
F	−.12	−.88*	−.28	+1.00*	+.75*	
G	+.29	−.62*	+.23	+.35	+.72*	+.61*

***Indicates a chi square significant at the .05 level.**

The three bureaucratic characteristics are all positively associated with one another. The four rational characteristics are likewise all positively associated. The general pattern of associations between sets, however, is negative; none of the three apparent exceptions is statistically significant.

Implications

These findings have implications for the use of the concept "informal" in organizational analysis, as well as for further model construction. Behaviorally, the concept "informal organization" has been used to refer to deviations from patterns described by a Weberian ideal type construct. The conceptual analysis presented here indicates that "informal organization" thought of in this way is an artifact which disappears when ideal types are abandoned in favor of a system of variables. The results strongly indicate, moreover, that it is a misleading artifact, since its apparent prevalence is enhanced by the ways in which the variables involved happen to be related. The major problem becomes one of discovering precise interrelationships among such variables, rather than comparing the characteristics of "informal" with "formal" organization.

On the other hand it should be noted that the concept "informal" has been used also in a different, normative, sense, to denote the status of activities not prescribed by official organizational rules. The present findings do not suggest condemnation of this usage, but they do raise some questions as to its application. The "official—unofficial" distinction not only cuts across the present analysis, but also crosscuts the original ideal type, strictly speaking, in that the content of official rules is logically independent of organizational behavior. Thus from the viewpoint of model construction this distinction is unstable, since behaviorally similar organizations can vary greatly in terms of the content of their official rules. Hence, although the "official—unofficial" distinction is unquestionably useful for certain restricted purposes, it may be highly misleading as a general theoretical orientation.

Serious questions, therefore, can be raised as to the long-run usefulness of the "formal—informal" dichotomy, except where it is used narrowly to mean "official—unofficial." And cautions are in order with respect to this latter usage in the construction of behaviorally centered models.[9]

As an alternative to the "formal—informal" dichotomy, the results of the present study suggest the core of a general model of formal organization, as stated by the following three propositions:

1. The technological nature of the task being performed determines the levels of both bureaucracy and rationality at which the organization must minimally operate.
2. Bureaucracy and rationality tend to be mutually inconsistent in the same formal organization.
3. In the face of such inconsistency, accommodative mechanisms arise which result in the continued operation of the organization at some level of efficiency.

The first proposition is based on findings reported and discussed elsewhere;[10] the second results from the present study. The third proposition is reached from the first two through a "functionalist" line of reasoning and demands further explanation. Given further knowledge of the actual processes involved in the presumed conflict between bureaucratic and rational elements, it should be possible to isolate "accommodative mechanisms" and to systematize their total range of variation. For the present, however, it is possible to give only more or less persuasive empirical examples which seem amenable to the interpretation suggested. Argyris and Gouldner, for instance, independently report that the combination of segmented, specialized work with a close, authoritarian supervisory style leads to interpersonal tensions among organization members. Gouldner concludes that the use of general, impersonal rules partially accommodates this conflict by lowering the visibility of bureaucratic power relations. Such accommodation, however, is incomplete; furthermore, imposition of the rules in itself is likely to involve tensions of its own. Other accommodative mechanisms thus may be expected; Argyris lists seven patterns which possibly are amenable to this interpretation: turnover, frenetic attempts to climb the organizational ladder, "goldbricking," rate setting, growth of cliques, development of personal defense mechanisms, and increased stress on material rewards for work.[11]

These examples—which could be readily multiplied—are intended merely to be illustrative. They do offer some evidence, however, that the model suggested is realistic and provides a unifying context for many apparently diverse phenomena which are commonly observed in organizational research. The principal variables which emerge in the system are: (1) the minimum level of bureaucracy, as technologically determined; (2) the minimum level of rationality, as technologically determined; (3) the degree of accommodation necessary between bureaucracy and rationality for some level of efficiency (4). The state of (1) and (2) determines (3) for given values of (4); (3) is presumably composed of several commensurable dimensions, each of which represents possible alternative patterns which are to some degree substitutable one for another.

It would seem, therefore, that Weber's original ideal type can be made to serve as a basis for the construction of a model which accounts for a much wider range of phenomena than Weber is generally credited as having considered—phenomena which frequently have been treated in ad hoc fashion as "informal" characteristics. Such a model, however, is more complex than might first appear, since empirical investigation reveals that a Weberian ideal-typical "rational bureaucracy" is likely to be unstable as a social system. Research is currently being directed toward a more operational specification of the variables suggested, as well as a more detailed explication of their interrelationships.

NOTES

1. See Talcott Parsons, *The Structure of Social Action,* Glencoe: Free Press, 1949, pp. 601–610.

2. Parsons, *loc. cit.*; also Carl J. Friedrich, "Some Observations on Weber's Analysis of Bureaucracy," in Robert K. Merton, *et al.*, editors, *Reader in Bureaucracy*, Glencoe: Free Press, 1952, pp. 27–33.

3. See, respectively, Wilbert E. Moore, *Industrial Relations and the Social Order*, New York: Macmillan, 1951, pp. 273–293; Peter M. Blau, *Bureaucracy in Modern Society*, New York: Random House, 1956, p. 57; James G. March and Herbert A. Simon, *Organizations*, New York: Wiley, 1958, pp. 34–82.

4. George Peter Murdock, "World Ethnographic Sample," *American Anthropologist*, 59 (August, 1957), pp. 664–687; George Peter Murdock, Clellan S. Ford, *et al.*, *Outline of Cultural Materials*, New Haven: Human Relations Area Files, 1950.

5. These variables, adapted from specifications alleged by Weber in various parts of

his work to apply to "rational bureaucracy," are not necessarily those conventionally cited in this connection, nor are they intended to exhaust all those given by him. See T. Parsons, editor, *Max Weber: The Theory of Social and Economic Organization* (translated by A. M. Henderson and T. Parsons), New York: Oxford, 1947, pp. 225–226; *From Max Weber: Essays in Sociology* (translated and edited by H. H. Gerth and C. W. Mills), New York: Oxford, 1946, pp. 196 ff.; *General Economic History* (translated by F. Knight), Glencoe: Free Press, 1950, pp. 95–96.

6. Organization charts were reconstructed from ethnographic descriptions, and the number of authority levels counted. In an earlier formulation, the concept "bureaucracy" was defined in this manner. But it has subsequently seemed desirable to give "bureaucracy" a broader denotation, as is done below. See Stanley H. Udy, Jr., " 'Bureaucratic' Elements in Organizations: Some Research Findings," *American Sociological Review*, 23 (August, 1958), pp. 415–418.

7. In strict Weberian terms, this staff *is* the "bureaucracy." It is treated here, however, simply as one of a number of possible formal organizational characteristics.

8. Helen Constas, "Max Weber's Two Conceptions of Bureaucracy," *American Journal of Sociology*, 52 (January, 1958), pp. 400–409.

9. For more comprehensive discussion of the problems of the concept "informal," see Henry W. Bruck, "The Concept 'Informal' in Organization Theory," paper read at the annual meeting of the Eastern Sociological Society, 1955.

10. Stanley H. Udy, Jr., *Organization of Work: A Comparative Analysis of Production among Nonindustrial Peoples*, New Haven: HRAF Press, 1959, pp. 36–54; also Udy, "The Structure of Authority in Nonindustrial Production Organizations," *American Journal of Sociology*, 54 (May, 1959), pp. 582–584.

11. Chris Argyris, *Personality and Organization*, New York: Harper, 1957, pp. 54–79, 118–119; and Alvin W. Gouldner, *Patterns of Industrial Bureaucracy*, Glencoe: Free Press, 1954, pp. 231–245. See also March and Simon, *op. cit.*, pp. 44–46; and William J. Goode and Irving Fowler, "Incentive Factors in a Low Morale Plant," *American Sociological Review*, 14 (October, 1949), pp. 618–624.

2

D. S. PUGH, D. J. HICKSON,
C. R. HININGS, AND C. TURNER

DIMENSIONS OF ORGANIZATION STRUCTURE

A major task of contemporary organization theory is the development of more sophisticated conceptual and methodological tools, particularly for dealing systematically with variations between organizations. Udy,[1] for example, feels that "comparative analysis is an appropriate initial, boundary-setting approach to general organizational theory." Without it, case studies remain haphazard and generalizations remain dubious. Mayntz[2] is of the same mind, but she fears that to regard all organizations as comparable systems is so abstract that "propositions which hold for such diverse phenomena as an army, a trade union, and an university, must necessarily be either so trivial or so abstract as to tell hardly anything about concrete reality." There is as yet insufficient evidence to support or refute this view.

This paper reports attempts to investigate and measure structural differences systematically across a large number of diverse work organizations, using scalable variables for multidimensional analysis. A previous paper[3] described the conceptual framework upon which the present studies are based, which accords closely with that which Evan[4] has advocated, and with concepts from which Hage[5] derives his axiomatic theory.

From an examination of the literature on organizations, six primary dimensions of organization structure were defined: (1) specialization, (2) standardization, (3) formalization, (4) centralization, (5) configuration, (6) flexibility. These "constitutive" definitions, as Kerlinger[6] has termed them, were then translated into operational definitions, and scales con-

Reprinted from D. S. Pugh, D. J. Hickson, C. R. Hinings, and C. Turner, "Dimensions of Organization Structure," *Administrative Science Quarterly*, 13 (June, 1968), pp. 65–91.

structed. Scales were also constructed for aspects of organizational context, and these were used as independent variables in a multivariate analysis to predict structural forms. The present paper describes the methods and results of scaling the structural variables.

Sample

Data were collected on 52 organizations in the Birmingham area. Of these, 46 were a random sample stratified, by size and product or purpose, according to the Standard Industrial Classification of the British Ministry of Labour. They include firms making motor cars and chocolate bars, municipal departments repairing roads and teaching arithmetic, large retail stores,

TABLE 2-1
Sample: Organizations Studied (n = 46)

Number of employees		
251-500	*501-2,000*	*2,001+*
Metal manufacture and metal goods		
Components	Metal goods	Nonferrous
Components	Metal goods	Metal automobile components
Research division	Metal goods	
Components	Domestic appliances	
Manufacture of engineering and electrical goods, vehicles		
Components	Engineering tools	Automobile components
Vehicles		
Components	Repairs for government department	Commercial vehicles
Components		Vehicles
	Automobile components	Carriages
	Engineering components	
Foods and chemicals, general manufacturing, construction		
Food	Civil engineering	Confectionery
Paper	Glass	Public civil engineering department
Toys	Printer	
Abrasives	Food	Brewery
		Automobile tires
Services: public, distributive, professional		
Government inspection department	Public water department	Public education department
	Department store	Public transport department
Public local savings bank	Chain of retail stores	Bus company
Public baths department	Chain of shoe repair stores	Cooperative chain of retail stores
Insurance company		

small insurance companies, and so on. These 46 organizations were distributed as shown in Table 2–1 (three organizations of the original sample felt unable to cooperate and were replaced in the present sample). Data were also available for the other six organizations. The information on the 52 organizations was used in the construction of scales, since the analysis was for internal consistency and scalability and because it formed a larger pool of data. For all analyses relating scales to each other or relating structural variables to contextual ones, only data on the 46-organization sample were used.

The sample of organizations was drawn from the 293 employing units in the area, which had more than 250 employees. Each was a unit listed as an employer by the British Ministry of Labour, irrespective of ownership, so that the sample includes several units that belong to the same industrial group or to the same local government. In ownership, the sample ranged from independent family-dominated firms to companies owned by private shareholders, a cooperative, branch factories of large organizations, municipal departments, and national organizations. Since these employing units are work organizations and pay their members, voluntary organizations were excluded.

Method

Organizations were first contacted by a letter addressed to the chief executive of the Birmingham organizational unit, who might be a works manager, an area superintendent, a chairman, or some other administrator. Field work began with interviewing him at length. There followed a series of interviews with department heads of varying status, as many as were necessary to obtain the information required. Interviews were conducted with standard schedules listing the data desired. Since these data were descriptive and about structure, not personal data about the respondent, no attempt was made to standardize interview procedure. Wherever possible, documentary evidence was sought to substantiate verbal accounts. Interviews took place between mid-1962 and mid-1964.

As the method used was that of wide-ranging interviews within a comparatively short time—from a few days to several weeks spent in an organization—it was not possible to obtain adequate data on the sixth variable, flexibility, which involves changes in structure and requires a more detailed study over a longer period of time.

It is the strength and the weakness of this project that no items were used unless they were applicable to *all* work organizations, whatever they did; several possible items of information had to be sacrificed to this end. Since the research strategy was to undertake a wide survey to set the guidelines, the result was superficiality and generality in the data. The project deals with what officially *should* be done, and what is in practice *allowed* to be done; it does not include what is *actually* done, that is, what "really" happens in the sense of behavior beyond that instituted in organizational forms.

It also avoids, or at least attempts to minimize, the employees' perceptions of their organizations. This is in contrast to the kind of data reported by Hall[7] on hierarchy of authority, division of labor, rules and procedures, impersonal relations, and selection by competence; and by Aiken and Hage[8] on centralization, both hierarchy of authority and participation, and formalization, which approximates the definition of standardization on this project. Both use scores from forced-choice responses of employees to subjective statements about work practices.

The data collected were analyzed under the headings of the conceptual scheme. Scales were constructed to define the variables operationally. These measured the degree of a particular characteristic present by linking together a large number of items that could show this characteristic. The basic methodological problem to be faced was whether the results on single items could be added up to form, if not an equal interval dimension (such as height or weight), at least a stable ordered scale (such as intelligence or neuroticism) to represent the characteristic.[9] Such scales make it possible to undertake correlational analyses with subsequent multivariate prediction. The example of intelligence is apposite, since the procedures for statistical analysis have drawn primarily upon the methods of psychological test construction.

This problem was approached by carrying out item analysis of the data on a particular variable, using the Brogden–Clemans coefficient[10] to test whether the items scaled and could therefore be regarded as representing a dimension. The advantage of using this relatively little-known coefficient as an index of item-total correlation is that no assumption is required about the underlying score distribution, as would be the case for other indices.[11] Indeed, as Lord has shown, even when a distributional assumption can be made, this coefficient gives a better estimate of correlation than does an index that makes such an assumption.[12] The item analysis values show the extent to which the actual distribution of a particular item corresponds to the distribution, assuming that all the items formed a perfect scale in terms of cumulative scaling procedures. An item analysis value of 1.00 would indicate that the actual and perfect distributions are the same. This is seldom obtained with the data discussed here. A judgment has to be made as to whether a variable can be represented by a score on a scale, taking into account the size of the item analysis values, the number of items going into the scale, and the further analysis to which the scale is subjected. Multivariate procedures combining several scales, to form profiles for example, make lower mean item analysis values acceptable. Principal components analysis[13] can then be used in the identification of underlying factors and, in many cases, these can be conceptualized as variables summarizing a number of scales in the original list.

This procedure was followed in constructing 64 scales. Some of these were major dimensions, such as overall formalization or centralization; some were subscales concerned with only parts of a major variable, such as formalization of role definition, centralization of decisions affecting the whole organization; some were summary scales extracted by principal-components analysis to summarize a whole dimension, as with overall role specialization; or certain aspects of it, as with standardization of procedures defining task and image. The scales comprising configuration, the "shape" of the organization's role structure, were concerned with totals and percentages, which present no problems of scaling. The scales are listed in Table 2-2.

The objective was to assemble items that were as representative as possible of the potential population of such items in a work organization. For example, if specialization scales are to measure the specialization of functions and roles in an organization, then the activities included as items in the scales must represent fully all activities engaged in by an organization; standardization scales must include a representative sample of the procedures that could feasibly be used; and so on. The generalized description of the activities of organizations given by Bakke[14] was used to guide the search for items, for although very abstract, it does point to every range of activities. For example, the danger that standardization is measured only with items on procedures regulating *workflow* activities (such as procedures regulating production work) is avoided if attention is drawn also to *perpetuation* procedures (such as those about buying materials or engaging employees) and to *control* procedures (quality inspection, budgeting, etc.). Bakke's concepts also have the merit of being applicable to every work organization, whether industrial, commercial, retail, or otherwise.

TABLE 2-2
Dimensions and Scales

Scale number of dimension	*Scale title*
Specialization	
51.01	Functional specialization
51.02-51.17	Specialization no. 1-16
51.18	Qualifications
51.19	Overall role specialization (51.02-51.15, and 51.17)
Standardization	
52.00	Overall standardization
52.01	Procedures defining task and image
52.02	Procedures controlling selection, advancement, etc.
Formalization	
53.00	Overall formalization
53.01	Role definition
53.02	Information passing
53.03	Recording of role performance
Centralization	
54.00	Overall centralization of decisions
	Criteria to evaluate performance:
54.01	Finance
54.02	Costs
54.03	Time
54.04	Quality
54.05	Labor relations
54.06	Output volume
54.07	Decisions affecting whole organization
54.08	Decisions affecting subunits of organization
54.09	Decisions affecting individual
54.10	Autonomy of organization to make decisions
Configuration	
55.08	Chief executive's span of control
55.09	Subordinate ratio
55.42	Status of specializations
55.43	Vertical span (height) of workflow hierarchy
55.44	Direct workers (%)
55.46	Female direct workers (%)
55.47	Workflow superordinates (%)
55.48	Non-workflow personnel (%)
55.49	Clerks (%)
55.50-55.65	Size of specializations (%)
Traditionalism	
56.00	

By this means a list of items pertinent to each variable was prepared. The method was then to ask each organization whether it has a specialized role for each of the potentially specializable functions, or a standardized procedure for each of the standardizable routines, and so on. Since the potential population of such items is not known, strictly speaking no claim can be made as to the representativeness of the scales; but the fact that there are internal consistencies among widely ranging items that make it possible to construct scales, supports the view that stable characteristics of the kinds conceptualized are being measured.

Primary Dimensions of Structure

SPECIALIZATION

Specialization[15] is concerned with the division of labor within the organization, the distribution of official duties among a number of positions.[16] Analysis of data from a pilot survey of organizations in terms of the Bakke activity variables made it possible to construct a list of sixteen activities that are assumed to be present in *all* work organizations, and on which any work organization may therefore be compared with any other. These activities or functions exclude the workflow activities of organization, and so are not concerned with operatives in manufacturing, sales clerks in retailing, and similar activities. Thus it can be seen whether an activity is specialized in an organization; that is, performed by someone with that function and no other, who is not in the workflow superordinate hierarchy (line chain of command). An endorsement of a specialization in this scale by an organization means only that the particular specialization is performed by one or more persons full time. No account is taken here of the *number* of specialists (this is an aspect of configuration, as discussed below) or of their status. Furthermore, only functions carried out by the organization itself are included in this scale: use of specialists from other organizations, for example, consultants, service agencies, experts from the head office, is considered to be an aspect of the organization's interdependence with its context.

Item analysis carried out using the Brogden–Clemans coefficient gave a mean item analysis of 0.76, which seems satisfactory for a first attempt. It was thus meaningful to talk of a scale of Functional Specialization and of an organization's score on it. The organizations ranged from those in which all these sixteen activities were performed by nonspecialists, to those in which they were all performed by specialists working in a functional relationship to the workflow management.

A second aspect of specialization is the extent to which specialist roles exist within each of the sixteen functional specializations, that is, role specialization. In those organizations with *functional specialization*, data were collected on the distribution of subtasks, and by using the Brogden–Clemans coefficient, a subscale covering each specialization was developed.

A further development of this analysis was to examine the relationship among all the role specializations. A principal-components analysis of all sixteen role-specialization scales extracted a first factor accounting for 65 percent of the variance, which is heavily loaded (range of 0.59 to 0.88) on all except one of the scales—legal specialization (scale no. 51.16) being the exception. On this basis, a scale of Overall Role Specialization (scale no. 51.19) was formed by summing the scores on the remaining fifteen scales. An organization could now be meaningfully characterized by its degree of specialization of role on a scale ranging from 0, for a government inspection agency in which no specializations, as defined here, were found; to 87, representing the highest degree of role specialization occurring in the sample, for a large confectionery manufacturing company. This is a much wider range of scores and therefore gives much finer discrimination than the original sixteen-point functional specialization scale, and in a single scale operationalizes the constitutive concept of specialization.

STANDARDIZATION

Standardization of procedures is a basic aspect of organizational structure, and in Weber's terms would distinguish bureaucratic and traditional organizations from charismatic ones. The operational problems here revolve around defining a procedure and specifying which procedures in an organization are to be investigated. A procedure is taken to be an event that has regularity of occurrence and is legitimized by the organization. There are rules or definitions that purport to cover all circumstances and that apply invariably.[17] The score is obtained by a count of the number of such procedures available to an organization from those in a given list. No assumption is made as to the use of procedures. Scores obtained ranged from 30, by a chain of retail stores, to 131, by a metals processing plant. This scale has a relatively low mean item analysis value of 0.41, partly because of the large number (157) of heterogeneous items used. However, when a principal-components analysis of scores on the items was carried out, it produced two meaningful factors. One factor, accounting for 21 percent of the variance (after correcting for negative roots) was defined as Procedures Defining Task and Image. The 33 items with a loading of 0.5 and above on this factor were collected into a scale giving a mean item analysis value of 0.65. The second factor, accounting for 13.5 percent of the variance, was a bipolar one. The positive end of the factor was concerned with the standardization of Procedures Controlling Personnel Selection and Advancement, and so on; the negative end of the factor was nonstandardization of workflow control. A scale formed from the items with a loading of 0.5 and over on the second factor (reversing the scores where the loading was negative) gave a mean item analysis value of 0.62. A high score on this scale therefore means not only that the organization *does* standardize its procedures for selection and advancement, and so on, but also that it *does not* standardize its procedures for workflow control.

FORMALIZATION

Formalization denotes the extent to which rules, procedures, instructions, and communications are written.[18] Definitions of 38 documents were assembled, each of which could be used by any known work organization. Adding an assessment of the range of personnel to whom a document applied, gave a total of 55 items and sub-items. A *document* is at least a single sheet of paper; therefore several copies of the same sheet of paper may each score as separate documents if used for separate purposes. For example, organization A may score 3 for unrelated pieces of paper, while organization *B* may score 3 for a set of carbon copies, each of which is detached for a different purpose. The problem of a single sheet of paper serving separate purposes did not arise. The mean item analysis value of the 55-item scale was 0.63. Scores ranged from 4, for a single-product foodstuffs factory, to 49, for the same metals-processing plant that headed the standardization scores.

A principal-components analysis of all the items gave a first factor, taking out 34 percent of the variance, but not extracting any factors meaningfully distinctive from overall formalization. Nevertheless, it was felt to be conceptually desirable to make some distinctions within the overall scale, which was split into three subscales concerned with Formalization of Role Definition, Information Passing, and Recording of Role Performance. The documents grouped together to constitute items on the subscale of formalization of role definition were all those designed primarily as prescriptions of behavior; for example, written terms of reference, job descriptions, and manuals of procedures. Information-passing documents were those intended to pass from hand to hand; for example, memo forms, and house journals. Role-performance records notified or authorized in writing the accomplishment of some part of a role; for example, records of the carrying out of inspection, or of maintenance of equipment.

The items going into these subscales gave mean item analysis values of 0.74 for Role Definition, 0.68 for Information Passing, and 0.67 for Recording of Role Performance. Unlike subscales formed to represent orthogonal factors (for example, standardization), these subscales are highly intercorrelated, correlations ranging from 0.55 to 0.81.

Although filing is characteristic of a bureaucracy, an attempt to discriminate between organizations in this respect failed. Once a document is in existence, copies of it appear to be filed for a very long time.

CENTRALIZATION

Centralization has to do with the locus of authority to make decisions affecting the organization.[19] Authority to make decisions was defined and ascertained by asking, "Who is the last person whose assent must be obtained before *legitimate action* is taken—even if others have subsequently to confirm the decision?" This identified the level in the hierarchy where executive action could be authorized, even if this remained subject to a routine confirmation later, for example by a chairman or a committee. A standard list of 37 recurrent decisions was prepared covering a range of organizational activities. For each organization, the lowest level in the hierarchy with the formal authority to make each decision was determined. Table 2–3 gives a generalized paradigm, which makes it possible to compare the levels across organizations. This overcomes the problem of deciding whether a foreman in factory A is at the same level as a shop-buyer in retail store *B* or a head clerk in commercial office *C*. Levels were equated in terms of the scope of the segment of workflow (that is, the proportion of the production activities) that they control. Where authority does not ultimately reside in the workflow hierarchy (line management) but in a

TABLE 2-3
Centralization: Levels in the Hierarchy

		Examples		
Score	*Level*	*Metal manufacturer*	*Chain of retail shoe repair shops*	*Local education department*
5	Above chief executive	Board of group	—	City council
4	Whole organization	Managing director	Chairman	Chief education officer
3	All workflow activities	Production manager	Sales manager	Assistant education officer
2	Workflow subunit	Plant manager	Area manager	Headmaster
1	Supervisory	Foreman	Shop manager	Head of department
0	Operating	Direct worker	Repairer	Teacher

staff or service department, a judgment has to be made approximating this to the equivalent level in the hierarchy, but there are comparatively few instances of this. An organization may have more than one managerial grade at a given comparative level; or it may not have a grade corresponding to a particular level.

The 37 decisions generated 148 biserial items. Eliminating items with poor discrimination gave a scale of 128 biserial items, which has a mean item analysis value of 0.40. Scores were obtained by scoring a decision taken outside the unit of organization (for example, at the head office) as 5, a decision taken at the chief executive level as 4, and so on to 0 for a decision at the operating level. Thus a high score means highly centralized, the maximum possible score being 128. The scores range from the most decentralized organization with a score of 51, an independent manufacturer of transporting equipment, to an extremely centralized organization with a score of 116, a branch factory where most of the decisions were taken right at the top or above the chief executive.

A further analysis of centralization scores indicated the degree of autonomy of the particular organizational unit. How many decisions did it have to refer to a headquarters or to a parent organization? This varied from independent companies, where the owner managers had complete control over all the operations of the enterprise, to a government agency, where a considerable number of decisions had to be referred upwards to higher authority. A scale was constructed by selecting from the 37 decisions of the main centralization scale those 23 decisions that showed the greatest discrimination between organizations as to whether the decisions were taken inside or outside the organization. The total score of an organization was the number of the 23 decisions that lay within its jurisdiction. A high score means great autonomy, and the range was found to be from 1 to 23. The mean item analysis value of the scale is 0.74.

CONFIGURATION

Configuration is the "shape" of the role structure.[20] Its data would be contained in a comprehensive and detailed organization chart that included literally every role in the organization. The assessment of the configuration of this hypothetical chart requires the use of a combination of selected dimensions, each of which provides a measure of the development of a particular aspect of the structure.

The vertical span of control (or height) of the workflow superordinate hierarchy (line chain of command) was measured by a count of the number of job positions between the chief executive and the employees directly working on the output. Lateral "widths" could include the chief executive's span of control, the ratio of subordinates to first workflow superordinates (first-line supervisors), and the percentage of total employees that were direct-output employees. Note that in a savings bank, the cashiers are direct-output employees, as are drivers in a bus company, and so on. The total workflow employees, that is, those directly responsible for the output (including management), could be compared with the number of those engaged in other activities, functional specialists, or staff departments. Totals of employees in each of the sixteen specializations could be calculated and related. Here the number of those engaged in a specialization was taken account of, apart from their specialization, which was measured with the specialization scale.

For example, a symptom of bureaucratization may be the number of clerks. A clerical job is defined as one where the main prescribed task is writing and recording, but where there is no supervisory responsibility for subordinates other than typists. This definition excludes office managers with authority over other clerks, but would include both a clerk and his typist. For the 52 organizations, the percentage of clerks ranged from 2 percent in small manufacturing firms to 29 percent in a commercial office.

TRADITIONALISM

An ideal standardization scale would be composed of items adequately representing a potential population of *customs* in organizations (a custom is an implicitly legitimized verbally transmitted procedure) and a potential population of *bureaucratic procedures*, being those explicitly legitimized by commitment to written form in rules, instructions, and other forms. Such a scale would indicate the extent to which an organization was standardized by customs or by rules. If rules were the means of standardization used, the formalization scores of the organization would be far higher than those of an *equally* standardized organization where the customs prevailed, since the rules would be embodied in documents. The relation between standardization scores and formalization scores could therefore be held to illustrate the Weberian distinction between traditional and bureaucratic types.

Because of the comparatively brief acquaintance of the research workers with the organizations studied in this project, the scores on the standardization scales reported here are thought to represent the rules and instructions of bureaucracy rather than procedures of customs. Nevertheless, ten items in the standardization scales were selected, which roughly corresponded to document items in formalization scales; for example, comparing routine frequency of inspection with documents recording the inspection, and central interviewing procedure with employment application forms. The reliability of field-work methods was supported when it was confirmed that organizations scored higher on the ten standardization items than on the equivalent ten formalization items: the reverse would have purported to show documents but not the procedures whose prescription or operation the documents recorded. The assumption was then made that if an organization had a procedure but did not have a corresponding document, then the procedure was more customary than bureaucratic. The hypothesis was that the degree to which an organization's standardization score on these items exceeded its formalization score reflected in a crude way its Traditionalism. An index was constructed, which ordered organizations on this discrepancy in scores.

The highest score was obtained by an old, established, small, branch factory, while the lowest point on the scale included a number of large manufacturers, a research organization, and an omnibus company. The association with the patriarchal–traditional type is direct, even though a question must remain about the adequacy of this simple scale as a measure of the type.

Intercorrelation of Structural Variables

STANDARD SCORES

The measures described afford comparisons between organizations on any one scale but not on scales of different variables, such as a comparison of an organization's standardization score with its centralization score. Comparability is obtained by converting raw scores into standard scores with a common mean of 50 and standard deviation of 15. This makes it possible to set the many scores of an organization side by side as a *profile* of its structure. Figure 2–1 shows the profiles of a number of organizations.[21]

INTERCORRELATIONS OF STRUCTURAL SCALES

With scales to represent the postulated primary dimensions of structure, one can explore the relationships between the dimensions and search for underlying similarities.

Intercorrelating the 64 scales of structure variables produces a large matrix of 2,016 coefficients. Selecting the scales that most fully represent

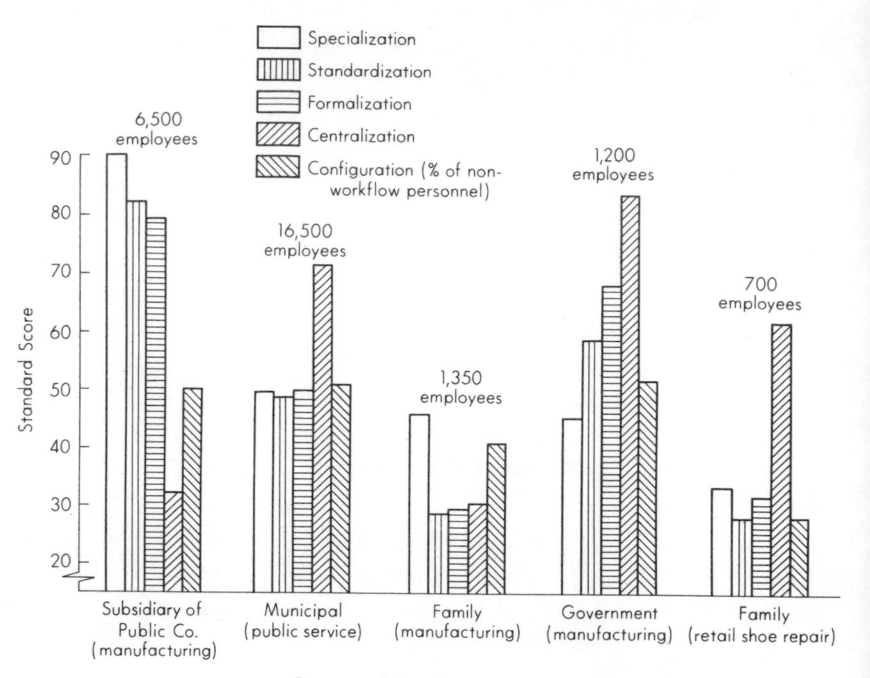

FIGURE 2–1 Structural Profiles of Five Organizations.

the variables and that were most distinctive gave the small matrix in Table 2–4. It is evident that many of the measures of structure are highly intercorrelated. For example, Overall Role Specialization (scale no. 51.19) has a correlation of 0.80 with Overall Standardization (scale no. 5200), of 0.68 with Overall Formalization (scale no. 53.00), of 0.66 with Configuration, Vertical Span (scale no. 55.43), and 0.56 with Configuration, Percentage Non-Workflow Personnel (scale no. 55.48). In other words, an organization with many specialists tends also to have more standard routines, more documentation, and a larger supportive hierarchy. A hypothesis on this process is that as specialists increase in number, they introduce procedures to regulate the activities for which they are responsible —the personnel specialist his selection procedure, the inspector his quality control—resulting in documentation—the application forms for vacancies and the inspection reports. A tall hierarchy results to encompass the specialists and the large number of non-workflow jobs.

This intercorrelation of these five scales contrasts markedly with the correlations between Centralization (scale no. 54.00) and these scales, which are all *negative*, and much smaller. This appears to disprove the

TABLE 2-4

Product-moment Correlation between Selected Scales of Structure (n = 46)

No.	*Title*	*Functional specialization*	*Legal specialization*	Overall *role specialization*	Overall *standardization*	*Standardization—selection, etc.*	Overall *formalization*	*Recording of role performance*	Overall *centralization*	*Autonomy of organization*	*Chief executive's span*	*Subordinate ratio*	*Vertical span (height)*	*Workflow superordinates (%)*	*Non-workflow personnel (%)*	*Clerks (%)*	*Traditionalism*
51.01	Functional specialization	–															
51.16	Legal specialization	0.32	–														
51.19	*Overall* role specialization	0.87	0.34	–													
52.00	*Overall* standardization	0.76	0.27	0.30	–												
52.02	Standardization—selection, etc.	−0.15	0.47	0.09	0.23	–											
53.00	*Overall* formalization	0.57	0.26	0.68	0.83	0.38	–										
53.03	Recording of role performance	0.66	0.11	0.54	0.72	−0.12	0.75	–									
54.00	*Overall* centralization	−0.64	−0.04	−0.53	−0.27	0.30	−0.20	−0.27	–								
54.10	Autonomy of organization	0.50	−0.15	0.40	0.06	−0.52	−0.02	0.10	−0.79	–							
55.08	Chief executive's span	0.22	0.15	0.34	0.28	0.04	0.32	0.32	0.10	0.02	–						
55.09	Subordinate ratio	0.25	−0.14	0.05	0.13	−0.46	0.04	0.39	−0.14	−0.14	−0.16	–					
55.43	Vertical span (height)	0.57	0.48	0.66	0.57	0.23	0.48	0.33	−0.28	−0.06	0.24	−0.05	–				
55.47	Workflow superordinates (%)	−0.53	0.21	−0.33	−0.37	0.39	−0.24	−0.52	0.52	0.47	0.12	−0.50	−0.01	–			
55.48	Non-workflow personnel (%)	0.58	0.11	0.56	0.51	−0.02	0.46	0.43	−0.40	−0.32	0.10	0.01	0.21	−0.43	–		
55.49	Clerks (%)	0.17	0.12	0.29	0.31	0.31	0.29	0.08	−0.04	−0.05	0.12	−0.24	−0.01	−0.05	0.46	–	
56.00	Traditionalism	−0.36	−0.13	−0.26	−0.24	0.06	−0.47	−0.54	0.39	0.30	−0.22	−0.17	−0.14	0.19	−0.26	−0.08	–

hypothesis,[22] drawn from the Weberian tradition and the notion that bureaucracies pass decisions to upper levels, that specialization, formalization, and centralization would be highly positively correlated. Hage, too, incorporates in his axiomatic theory the proposition that "the higher the centralization, the higher the formalization" (standardization).[23] On the other hand, the correlations support his proposition summarizing Thompson, Gouldner, and Blau and Scott: "The higher the complexity, the lower the centralization,"[24] "complexity" for Hage being specialization of tasks plus the length of training required for the task.

Standardization of Selection Procedures (scale no. 52.02) is even more interesting. First, it breaks the pattern of a strong positive relationship within specialization, standardization, formalization, and many configuration variables, correlating only 0.09 with Overall Role Specialization (scale no. 51.19) and 0.37 with Overall Formalization (scale no. 53.00). The low correlation of scale no. 52.02 with the other standardization scales would be anticipated, because it is constructed to represent an independent standardization factor, and it therefore suggests a structure of a different kind. Second, all its correlations with centralization are *positive*, the only scale of specialization, standardization, and formalization where this is so. In this the hypothesis of a positive relationship between standardization and centralization *is* supported, but applies to this aspect of standardization only.

The meaning of patterns of structural scores should be kept in mind when interpreting these correlations. An organization that scores high on specialization, standardization, and formalization (which is probable in view of the intercorrelations among these variables) would have gone a long way in the regulation of the work of its employees. As an organization, it would have gone a long way in *structuring* its activities; that is, the intended behavior of employees has been structured by the specification of their specialized roles, the procedures they are to follow in carrying out those roles, and the documentation of what they have to do. In short, what these three associated variables are exploring is the range and pattern of *structuring*.

The scales of centralization cannot be regarded as measures of structuring in this way. To assess structuring in terms of centralization would require measurement of how specific the loci of authority are; that is, how definite it is that authority for decision X rests in role Y. But the centralization scales treat this as a constant and measure only the vertical distribution of authority over the hierarchy. That one organization scores as centralized, whereas another scores as decentralized, does not necessarily bear any relationship to how specific the allocation of authority is within the two. Basing the measures of centralization upon a conceptual basis different from the rest of the structural variables has a very important effect upon the results.

UNDERLYING DIMENSIONS OF STRUCTURE

The intercorrelations suggested that the interpretation would be improved by a search for basic dimensions of organizational structure by the method of factor analysis. Principal-components analysis was applied to the matrix given in Table 2–4, and four factors were extracted, accounting in turn for 33, 19, 14, and 8 percent of the variance after rotation. These factors are orthogonal to one another, that is, mutually independent. The loadings of the variables on the factors are given in Table 2–5.

TABLE 2-5
Principal Components Analysis of Selected Scales of Structure after Graphic Rotation*

Scale No.	Scale Title	Factor I' Structuring of activities	Factor II'' Concentration of authority	Factor III' Line control of workflow	Factor IV Relative size of supportive component
52.00	Standardization	0.89*	−0.01	−0.21	0.10
51.19	Role specialization	0.87*	−0.33	0.01	−0.13
53.00	Formalization	0.87*	0.14	−0.21	0.17
56.00	Traditionalism	−0.41‡	0.18	0.32	−0.02
55.08	Chief executive's span	0.12‡	0.23	−0.07	−0.03
56.01	Functional specialization	0.78*	−0.47‡	−0.21	−0.17
55.48	Non-workflow personnel (%)	0.58†	−0.43‡	0.06	0.41‡
51.16	Legal specialization	0.51†	0.25	0.31	0.43‡
55.43	Vertical span	0.09*	0.03	0.08	−0.54†
55.49	Clerks (%)	0.40‡	−0.09	0.42‡	0.67*
53.03	Recording of role performance	0.69*	−0.05	−0.64*	0.13
55.09	Subordinate ratio	−0.05	−0.19	−0.80*	−0.06
52.02	Standardization—selection, etc.	0.40‡	0.59†	0.50†	0.09
55.47	Workflow superordinates (%)	−0.23	0.60*	0.50†	−0.22
54.00	Centralization	−0.33	0.83*	0.01	0.21
54.10	Autonomy of the organization	0.10	−0.92*	0.00	−0.13
	Variance (%)	33.06	18.47	12.96	8.20

*Weightings > 0.06.
†Weightings > 0.5.
‡Weightings > 0.4.

The meaning of the factors is readily apparent. Factor I can be seen to be most highly loaded on the variables of standardization, specialization, and formalization, as would be expected from the pattern in Table 2–4. This dimension therefore is called *structuring of activities*. The concept of the structuring of organizational activities has the advantage of applying to any or all parts of the organization; whereas it is a question whether the Weberian concept of bureaucracy can or cannot be applied outside the administrative hierarchy to the workflow operatives. Both clerical activities and shop floor activities can be more or less structured; whether they can both be bureaucratized is open to question.[25] Structuring therefore includes and goes beyond the usage of the term bureaucracy. It has the further advantage of being conceived and defined as an operationalized dimension, and not as an abstract type.

This usage accords with Etzioni,[26] who prefers to avoid terms such as bureaucracy, formal organization, and institution in favor of using the term organization "to refer to planned units, deliberately structured." Etzioni does not go on to use this definition to examine organization theory from the standpoint of variations in *structuring*, but this approach follows from Hickson's[27] argument that a common thread of this kind runs through the subject, which he calls the varying "specificity of role prescription." Specificity has a very close affinity to structuring. Thus standardization may be considered as the specificity of procedures to cases, specialization as the specificity of tasks to roles, and so on. Hickson's linking of the approaches of twenty-two organization theorists by the concept of specificity also suggests that measures of structuring are prerequisite to further development in the subject.

Factor II is marked by the opposition of centralization and autonomy and is thus concerned with *concentration of authority*. Specialization is in the direction of dispersed authority, as would be expected; with more specialization, authority is likely to be distributed to the specialists. "Non-workflow personnel" also has a definite weighting toward dispersed authority, but the percentage of workflow superordinates is firmly weighted in the direction of concentrated authority. Thus the greater the percentage of the line hierarchy, the more concentrated is the authority, and the less decentralized are the decisions down the line. Standardization of selection procedures is also positively loaded on this factor, as would be anticipated from its correlation with centralization.

In Factors III and IV, variables of configuration predominate. The third factor is characterized by the heavy positive loading of Percentage of

Workflow Superordinates and negative loading of Subordinate Ratio, and is concerned with the line hierarchy. Formalization of role performance records has a high negative loading, and inspection of the items of this scale (for example, record of work, record of time, record of maintenance) indicates that these records are those prepared on the workflow personnel for control purposes. The factor may thus be considered as that of *line control of workflow* (rather than impersonal control). This characterization is supported by the significant positive loading of the scale, standardization of procedures for selection, advancement, and so on (scale no. 52.02), the bipolar scale, and a high score on it means not only that there *are* procedures for selection, and so on, but also that there are *not* procedures for controlling the workflow. Thus control would be expected to rest in the hands of the workflow personnel themselves and their line superordinates.

Factor IV has its high loadings on percentage of clerks and percentage of non-workflow personnel. The factor is thus concerned with the amount of activity auxiliary to the main workflow of the organization. Following Haas *et al.*,[28] the term *supportive component* is used in relation to these activities. This factor is distinguished from Factor III, which was concerned with the *control* of *workflow* activities, whereas Factor IV is concerned with the *amount* of *auxiliary* activities of a non-control kind (all the structuring and authority variables are negligibly loaded). There are many non-control supportive activities, such as clerical, transport, catering, and others.

Thus it must be presumed that there are a number of distinctive underlying dimensions of structure—this particular trial produces four. Since these are mutually independent, an organization's structure may display all these characteristics to a pronounced degree, or virtually none at all, or display some but not others. In so far as the original primary dimensions of structure, specialization, standardization, formalization, centralization, and configuration were drawn from a literature saturated with the Weberian view of bureaucracy, this multifactor result has immediate implications for what we have elsewhere called the Weberian stereotype.[29] It is demonstrated here that bureaucracy is *not* unitary, but that organizations may be bureaucratic in any of a number of ways. The force of Blau's criticism of the "ideal type" can now be appreciated: "If we modify the type in accordance with the empirical reality, it is no longer a pure type; and if we do not, it would become a meaningless construct."[30] The concept of *the* bureaucratic type is no longer useful.

Summary and Discussion

From an examination of the literature of organization theory, six primary dimensions of organization structure were postulated: specialization, standardization, formalization, centralization, configuration, and flexibility. The first five of these concepts were then operationalized by generating 64 component scales to measure various aspects of the primary dimensions.

Some of these component scales were primary dimensions themselves, such as Overall Formalization; some were subscales concerned with only parts of a major variable, such as Autonomy of the Organization—an aspect of centralization, Percentage of Non-Workflow Personnel—an aspect of configuration, and so on. Some were summary scales extracted by principal-components analysis to summarize a whole dimension, as with Overall Role Specialization, or certain aspects of it, as with Standardization of Procedures for Selection and Advancement.

Comparative data across 52 different work organizations made it possible to test the internal consistency of these 64 scales and to examine the intercorrelations between them. A principal-components analysis of 16 of the 64 scales, which most fully represented the primary dimensions, yielded four empirically established underlying dimensions of organization structure: *structuring of activities*, encompassing Standardization, Formalization, Specialization, and Vertical Span; *concentration of authority*, encompassing Organizational Autonomy, Centralization, Percentage of Workflow Superordinates, and Standardization of Procedures for Selection and Advancement; *line control of workflow*, encompassing Subordinate Ratio, Formalization of Role Performance Recording, Percentage of Workflow Superordinates, and Standardization of Procedures for Selection and Advancement; and *relative size of supportive component*, encompassing Percentage of Clerks, Vertical Span, and Percentage of Non-Workflow Personnel.

The establishment of these scales and dimensions makes it possible to compile profiles characteristic of particular organizations, and examples of these are given. As a result of this dimensional analysis, it is clear that to talk in terms of the bureaucratic ideal type is not adequate, since the structure of an organization may vary along any of these four empirical dimensions. Figure 2–2 gives the profiles of the five organizations shown in Figure 2–1, but using the factor dimensions. It will be seen that these profiles pinpoint more sharply the structural differences between these

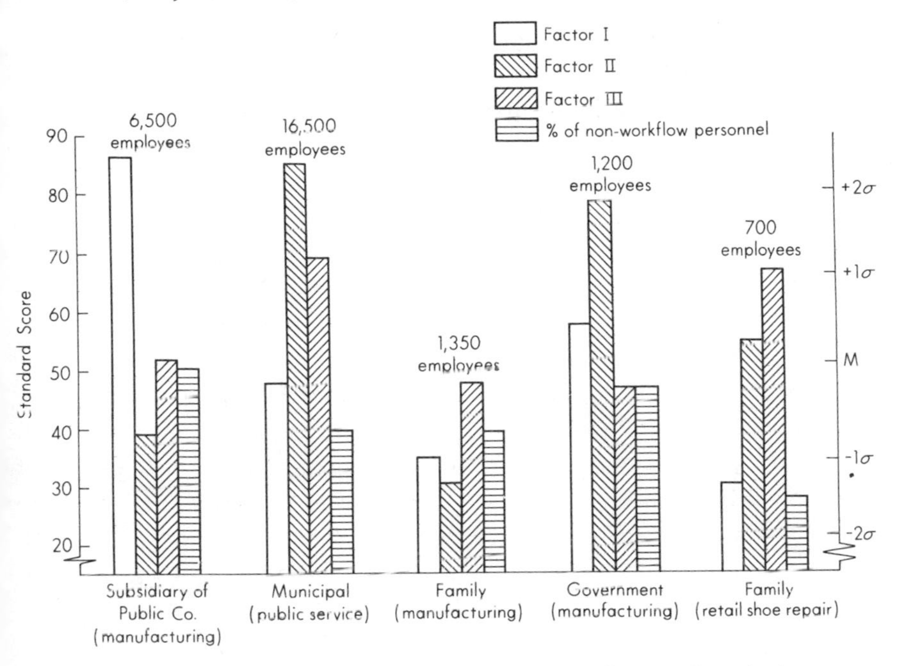

FIGURE 2–2 Underlying Dimensions of Structure in the Five Organizations in Figure 2–1.

organizations. Thus the subsidiary manufacturing organization is distinctively high on Factor I, Structuring of Activities. The municipal public service is high on Factors II, Concentration of Authority, and III, Line Control of Workflow; the family manufacturing is low on all factors; the government manufacturing is high on Factors I, Structuring of Activities, and II, Concentration of Authority; and the family retailing is distinctively high on Factor III, Line Control of Workflow.

The dimensional approach described here has immediate implications for further work. By making it possible to pinpoint structural differences between organizations, it makes it possible to conduct more systematic and rigorous studies of group composition and interaction, comparative role analysis of conflict and performance, and individual personality and behavior, since the effects of structural aspects of the organizations can be controlled.

The second major gain from establishing dimensions is that it makes possible a multivariate approach to causality. If similar scales can be developed for aspects of an organization's context, then the relationships between dimensions of context and dimensions of structure can be examined using correlational and multivariate techniques. This obviates the need to select a particular aspect of context *a priori* as the determining variable for structure. For example, both the variables of size and technology have been presented as *the* determining ones for structure, but the relative importance of these two factors has still to be demonstrated. A multivariate dimensional approach could attempt this.

The relative status of the primary conceptual variables, the subscales, and the empirically established underlying dimensions of structure have become clear. The conceptual dimensions and subscales have been demonstrated to be empirically meaningful, and hypotheses can be set up and investigated relating these aspects of structure to other aspects of the organization's context and functioning. The underlying dimensions have been shown empirically to be a parsimonious method of summarizing an organization's structure, and more suitable for many purposes of prediction and statistical control. In particular, the development of a taxonomic classification could be expected to be founded more stably on the underlying dimensions. Accepting that there are a number of component dimensions in structure, four of which are (1) structuring of activities, (2) concentration of authority, (3) line control of workflow, and (4) relative size of supportive component, gives an empirical basis for a taxonomy of organization structures.*

* Editors' Note: Appendices A–F appear at this point in the original article. Each appendix indicates the manner in which the scales were formed and the items making up the scales. Space limitations preclude their reproduction here. The inclusion of such information in journal articles is, however, strongly recommended.

NOTES

1. S. H. Udy, Jr., "The Comparative Analysis of Organization," in J. G. March (ed.), *Handbook of Organizations*, Chicago: Rand McNally, 1965.
2. Renate Mayntz, The Study of Organizations: a Trend Report and Bibliography, *Current Sociology*, 13 (1964), 3.
3. The authors are members of the Industrial Administration Research Unit, at the University of Aston in Birmingham, England. Research conducted by the Unit is jointly supported by the Social Science Research Council and the University. D. S. Pugh, D. J. Hickson, C. R. Hinings, K. M. Macdonald, C. Turner, and T. Lupton, A Conceptual Scheme for Organizational Analysis, *Administrative Science Quarterly*, 8 (December 1963), 289–315.
4. W. M. Evan, Indices of the Hierarchical Structure of Industrial Organizations, *Management Science*, 9 (April 1963), 468–477.
5. J. Hage, An Axiomatic Theory of Organizations, *Administrative Science Quarterly*, 10 (December 1965), 289–320.
6. F. Kerlinger, *Foundations of Behavioral Research* (New York: Holt, Rinehart and Winston, 1964).

7. R. H. Hall, The Concept of Bureaucracy: an Empirical Assessment, *American Journal of Sociology*, 69 (July 1963), 32–40; also Intraorganizational Structural Variation, *Administrative Science Quarterly*, 7 (December 1962), 295–308.

8. M. Aiken and J. Hage, Organizational Alienation: a Comparative Analysis, *American Sociological Review*, 31 (August 1966), 497–507.

9. C. H. Coombs, A *Theory of Data* (New York: Wiley, 1964).

10. H. E. Brogden, A New Coefficient: Applications to Biscrial Correlation and to Estimation of Selective Efficiency, *Psychometrica*, 14 (1949), 169–182.

11. We are grateful to Dr. P. M. Levy, Visiting Research Fellow, for bringing this statistic to our notice, and for much other statistical advice and encouragement.

12. F. M. Lord, Biserial Estimation of Correlation, *Psychometrica*, 28 (1963), 81–85.

13. H. H. Harmon, *Modern Factor Analysis* (Chicago: University of Chicago, 1960).

14. E. Wight Bakke, "Concept of the Social Organization," in M. Haire (ed.), *Modern Organization Theory* (New York: Wiley, 1959).

15. A preliminary discussion of this variable is given in C. R. Hinings, D. S. Pugh, D. J. Hickson, and C. Turner, An Approach to the Study of Bureaucracy, *Sociology*, 1 (January 1967), 61–72.

16. Pugh *et al., op. cit.*

17. Pugh *et al., op. cit.*

18. Pugh *et al., op. cit.*

19. Pugh *et al., op. cit.*

20. Pugh *et al., op. cit.*

21. A full report of the relationship between context and structure will be given in a later paper.

22. Pugh *et al., op. cit.*

23. Hage, *op. cit.* 300.

24. Hage, *op. cit.* 300.

25. For example, T. Caplow, *Principles of Organization* (New York: Harcourt, Brace, 1964), p. 287.

26. A. Etzioni, *Modern Organizations* (New Jersey: Prentice-Hall, 1964), p. 4.

27. D. J. Hickson, A Convergence in Organization Theory, *Administrative Science Quarterly*, 11 (September 1966), 224–237.

28. E. Haas, R. H. Hall, and N. J. Johnson, The Size of the Supportive Component in Organizations, *Social Forces*, 42 (1963), 9–17.

29. Hinings *et al., op. cit.*

30. P. M. Blau, Critical Remarks on Weber's Theory of Authority, *American Political Science Review*, 57 (June 1963), 305–316.

PART II

The Impact of External Factors on Organizations

Organizations, like individuals, are not "islands unto themselves." They are surrounded by and in interaction with other organizations, individuals, and the general social conditions of which they are a part. Like almost all social relations, the organization's interactions are reciprocal. They affect what they affect and what they are affected by.

Having stated what has become a cliché for interpersonal interactions and practically an article of faith in the analysis of organizations, the sad fact is that there is a real paucity of data regarding the manner in which external factors affect organizations. Part of the problem lies in the recentness of comparative work in which cross-sectional comparisons between organizations can be made on the basis of differing environmental circumstances. The general weakness in the ability to measure social processes is also clearly reflected in the absence of data on organizational-environmental transactions. Another factor which contributes to the state of affairs is that interest in external factors is strangely rather recent in the sociological literature. There are some exceptions to this, of course. Selznick's classic study of the Tennessee Valley Authority,[1] Clark's analyses of the California system of higher education,[2] Bendix's analysis of the conditions surrounding industrialization,[3] Abegglen's study of the impact of Japanese culture on factory work,[4] and Crozier's examination of two French bureaucracies[5] are important case studies which are very suggestive in terms of the strength and direction of environmental influences. Similarly Eisenstadt[6] and Stinchcombe[7] have paid particular attention to the importance of social structural characteristics for organizational development in their essays. The fact remains, however, that little in the way of systematic data is available for good comparative analyses.

Despite these problems there are some promising directions being suggested in the literature. A great deal of attention is being paid to the role of technology in contemporary organizational research. The term "technological determinist" has become an epithet in sociologists' jargon aimed at those who happen to believe in the overriding importance of technology. The major factors in the development of interest in this subject are that the technology variable is moderately easy to "get a handle on" and some important studies have been published which use this as the key variable.

Before looking at technology and other external factors it should be pointed out that technology will be treated here as an external factor. This is done because what an organization utilizes in its work (its technology) comes into the organization from the environment. Although a particular brand of fried chicken may have a "secret recipe," the general chicken-handling technology is shared among organizations operating in this area. In organizations having any contact at all with ideas developed through the research process, the interactions with and dependency on the "scientific community" are obvious. Ideas gleaned from the environment are largely outside the control of any one organization, so that the organization develops its technology on the basis of what it brings in from the environment.

The technology issue is confronted directly in the Hage and Aiken paper. It will be seen that there is strong support for the argument that the nature of the technology used by an organization is a major factor in determining the form it will take.

A premature sense of closure could be derived from this paper and the argument surrounding technology. Some research points to additional factors which seem to be as important as technology, while still other authors deny the importance of technology altogether. Blau and Schoenherr appear to attribute almost all structural differentiation to the varying sizes of the organizations in their research.[8] Pugh, *et al.*, conclude that size, together with technology and the organization's dependence on other organizations in its relevant environment are the major factors associated with the different forms of organizational structure found in their study. Clearly, the issue is not settled.

Further studies in the Pugh, *et al.*, mold will be extremely helpful in reducing the confusion about the relative importance of the various external elements under discussion. In a later paper, Hickson, *et al.*, suggest that the approach to technology would be improved by differentiating between operating, material, and knowledge technology.[9] This differentiation recognizes the fact that different parts of the organization actually use different technologies. The findings from this particular piece of research suggest that operations technology (used directly on the workflow) has an important impact on the structure of small organizations. It is in these organizations that the specific work being performed is extremely close to the balance of the organization not directly involved in the production or services performed. In larger organizations these other activities are insulated from the workflow and are hence less affected by this particular phase of the technological system. Data are unavailable regarding variations in the other forms of technology identified to determine if they are related to the structure of the balance of the organization which is not affected by operations technology. This sort of analysis will enable the researcher to sort out the effects of the several variables under investigation. The evidence is quite overwhelming, despite the absence of final evidence, that technology is strongly related to the form which an organization takes.

Organizational size is a factor which has received a great deal of attention in the literature. Although it is unclear whether to treat size as an external or an internal condition, we will follow Pugh, *et al.*, and discuss it briefly in this context. Some authors, including Chapin and Blau and Schoenherr, approach size as the major factor influencing the shape which organizations assume.[10] Others, including myself, have suggested that the impact of size is overridden by other more important variables, such as technology and the other factors to be discussed.[11]

While little is proven by assertion, the situation at present seems to be that size is related to organizational structure. Organizations are more formalized and more complex if they are large. At the same time the number of personnel and other resources devoted to administrative tasks tend to grow disproportionately as size increases, up to a point. After this point there appears to be a decline in the proportion of resources devoted to administrative tasks as organizational size increases further.

The evidence in regard to size is quite sound. The interpretations are the point of disagreement. In my opinion organizational size increases along with other structural changes as organizations face contingencies which force them to diversify into new areas, handle a larger population, or deal with new laws.[12] The decision is made within the organization to go into these new areas or whatever and the structural changes occur as a consequence of the decision that is made, rather than on the basis of simple increased size. Very few organizations would decide to increase their size for no reason. The reasons for the final decisions (and this does not make the rationality assumption) appear to be the major factors in determining both the increase in size and the more important changes in structure.

Much of the literature relevant to this section has been concerned with technology and size. This has led to overlooking other apparent important external conditions. Certainly demographic conditions are important for organizations as anyone with experience in the disposable diaper business or colleges and universities will attest. The simple passage or interpretation of laws makes a big difference to organizations involved with the particular laws in question. The general political climate is clearly important as historical analyses of the growth of organizations have shown and the contemporary state university is feeling. Economic conditions have an obvious, but generally ignored (at least by sociologists), impact on organizations. The conditions in the general environment have a clear impact on the organizations in that environment. This is an area in which there is very little good evidence, but one which is in great need of systematic investigation. Examinations of organizations in different cultural settings indicate the importance of this factor.[13] This is part of the organization's environment from which it cannot really escape.

If the relevant environmental factors can be identified, it may be possible over time to specify the relative strength and direction of environmental influences on organizations. It does not seem logical that all the factors discussed, here or in the articles that follow, would be equally important for organizations. Until more evidence is available, however, organizational research should continue to examine the variety of environmental interactions in which organizations constantly engage.

There is an additional component in an organization's external relations. This is the clients or other recipients of an organization's outputs. The Lefton and Rosengren article, one of the few dealing with this specific topic, attempts to show how the organization's forms of linkages with its clients affect its own structure and operations. These linkages are with relatively stable client groups. In the current context of client and student militancy (the possibility of consumer militance would present the same problems for the organization) the importance of clients for the organizations will become magnified for the organization and its members. This will be a direct impact on the organization from its environment. Aside from the theoretical interest in the impact of clients, there is a growing awareness among some sociologists that clients and other "recipients" of organizational outputs, including those living downstream or downwind from many organizations, have a claim that is increasingly being heard and legitimated on the organizations. This can take the form of legislation and thus be no different than other legislation or it can be in the form of protests, demonstrations, and other such pressures on the organization. These are facets of external influences on organizations on which research has barely begun to touch.

NOTES

1. Phillip Selznick, *TVA and the Grass Roots*, Berkeley: University of California Press, 1949.

2. Burton R. Clark, *Adult Education in Transition*, Berkeley: University of California Press, 1956; and *The Open Door College*, New York: McGraw-Hill, 1960.

3. Reinhard Bendix, *Work and Authority in Industry*, New York: John Wiley, 1956.

4. James C. Abegglen, *The Japanese Factory*, Glencoe: Free Press, 1958.

5. Michael Crozier, *The Bureaucratic Phenomenon*, Chicago. University of Chicago Press, 1964.

6. S. N. Eisenstadt, "Bureaucracy, Bureaucratization, and Debureaucratization," *Administrative Science Quarterly*, 4, no. 3 (December, 1959), pp. 302–320.

7. Arthur L. Stinchcombe, "Social Structure and Organizations," in James S. March, *Handbook of Organizations*, Chicago: Rand McNally, 1965, pp. 142–193.

8. Peter M. Blau and Richard R. Schoenherr (with Sheila R. Klatzky), *The Structure of Organizations*, New York: Basic Books, 1970.

9. David J. Hickson, D. S. Pugh, and Diana C. Pheysey, "Operation Technology and Organization Structure: An Empirical Reappraisal," *Administrative Science Quarterly*, 14, no. 3 (September, 1969), pp. 378–397.

10. F. Stuart Chapin, "The Growth of Bureaucracy: An Hypothesis," *American Sociological Review*, 16, no. 6 (December, 1951), pp. 835–856; and Blau and Schoenherr, *op. cit.*

11. Richard H. Hall, J. Eugene Haas, and Norman Johnson, "Organizational Size, Complexity and Formalization," *American Sociological Review*, 32, no. 6 (December, 1967), pp. 903–912.

12. This argument is strengthened by a re-analysis of the Pugh, *et. al.*, data by Howard Aldrich ("Theory Construction, Path Analysis, and the Study of Organizations: A Re-Examination of the Findings of the Aston Group," Unpublished manuscript: Cornell University, 1970). In this analysis size becomes a *dependent* variable, with dependence and technology playing a more important role. This point should be kept in mind when the Pugh paper is read.

13. See especially Abegglen, *op. cit.*, and Crozier, *op. cit.*

3

JERALD HAGE AND MICHAEL AIKEN

ROUTINE TECHNOLOGY, SOCIAL STRUCTURE, AND ORGANIZATION GOALS

In the last few years, technology has become increasingly popular as an explanatory concept in organizational analysis. Blauner (1964) used it as a key factor in explaining different levels of alienation in American industry. Woodward (1965) found that types of technology affected the number of levels of management and the span of control in British industry, and Perrow (1967) has suggested several hypotheses linking this concept to different aspects of the organization's structure and goals. The purpose of this paper is to explore these and other hypotheses in a study of people-processing organizations.*

This kind of organization provides a particularly interesting context for the study of the association between technology and the structure and functioning of an organization. Techniques of reinforcement or interviewing are less obvious than those of machine handling. Yet rehabilitation centers, mental hospitals, family agencies, and schools have a work-flow and add value to their product, the client, just as any industrial firm does. The difficult task is to find measurable dimensions for describing this work-flow. The terms used by Blauner (1964) and Woodward (1965)—continuous process or assembly-line—are clearly not applicable to people-

Reprinted from Jerald Hage and Michael Aiken, "Routine Technology, Social Structure, and Organization Goals," *Administrative Science Quarterly*, 14–3 (September, 1969), pp. 366–376.

* This investigation was supported in part by a research grant from Social and Rehabilitation Services, Department of Health, Education and Welfare, Washington, D.C. We are grateful to Charles Perrow for helpful comments on an earlier version of this paper. In addition, we would like to acknowledge the cooperation and support of Harry Sharp and the Wisconsin Survey Laboratory during the interviewing phase of this project.

processing organizations. Furthermore, these categories are also limited in industry in their application: they describe only some kinds of technologies. Therefore, one task is to locate a general technological dimension that can be used not only in industry but also in people-processing organizations as well.

Part of the difficulty in finding a general variable describing technology stems from the fact that the concept subsumes many different ideas. The construct is so rich that it would be better to treat it like that of the concept of social structure—one containing many dimensions that need to be delineated and measured. Litwak (1961) has suggested one basic dimension, the uniformity of tasks. Perrow (1967:195–196) has called attention to a similar idea, the routineness of work. However, he defines this as including both the relative stability or variability of the raw material as well as how much is understood about this material. If the clients are stable and uniform and much is known about the particular process of treatment, then the organization has a routine work-flow. The teaching of reading might be a good example. In the contrasting situation of little uniformity and little understanding, we have a nonroutine work-flow. The interviewing of a psychotic might be an apt illustration. Thus, Perrow's discussion of routine vs. nonroutine actually subsumes several dimensions.

The routineness of work does not cover all aspects of the concept of technology. There are other dimensions that can be defined and measured. For example, the environment of work, such as level of noise, dirt, and layout or the amount of energy used, is frequently included under the rubric "technology" (Labovitz, 1963). It is difficult, however, to apply this across all organizations. The degree of routineness is one dimension of technology that can be applied equally to people-processing, industrial, and other kinds of organizations, and it can provide the basis for general propositions that can be tested in many organizational contexts.

The interest in technology as an independent variable stems from the recognition that the work processes of an organization provide the foundation upon which social structure is built (Perrow, 1967:195). Because of this, technology should influence the nature of that structure. That is, technology is likely to determine whether it is formalized or nonformalized, whether it has a diverse or relatively simple division of labor, and so together, the technological foundation and substructural social arrangements influence the substructure of organizational goals. As Perrow has suggested, these factors set limits on the goals that are maximized and those that are minimized. It is this connection between routine work, organizational structure, and goals that forms the focus of this paper.

The intellectual problem posed here requires that the organization be treated as a unit of analysis and that properties of that unit—analytical, structured, and global (Lazarsfeld and Menzel, 1961)—be abstracted and analyzed. The routineness of the technology of the entire organization rather than the routineness of particular techniques is thus analyzed. Similarly, the concern with social structure is with the arrangement of all social positions or jobs in an organization, not just certain ones. The research design for sampling these arrangements and for measuring the complete organization as a unit of analysis are described in the next section. The specific hypotheses and findings relating the degree of routine work to structural variables and goals are then presented.

Research Design and Measurement

The data were collected in sixteen health and welfare agencies located in a large midwest metropolis during the early Spring of 1967. Ten agencies were private; six were either public agencies or branches of public agencies. These organizations were all the larger agencies that provide rehabilitation services, psychiatric services, and services for the mentally retarded as defined by the directory of the community chest. The organizations varied in size from just over twenty in a small family agency to over six hundred in one of the mental hospitals.

PROCEDURES FOR MEASURING ORGANIZATIONAL VARIABLES

A central problem in the study of organizations concerns the measurement of organizational properties or variables (Lazarfeld and Mensel, 1961; James Coleman, 1964). Organizations are composed of individuals working in various jobs which are arranged in different structural configurations and work-flow. It is these arrangements which will be described and measured in this paper.

The organization was divided into levels and departments and then job occupants were selected randomly within these categories. In other words, a stratified random sample in which the stratification was based on two key dimensions of the organization—levels and departments—was used. This procedure has several advantages: (1) it can be used in any kind of organization, irrespective of its specific goals; (2) it corresponds to the way organizations are actually structured; (3) organizational participants are likely to locate themselves in terms of different levels and departments. These criteria are not only analytical justifications, but also correspond to the reality of organizational life. Certainly the nature of work, an essential part of the problem of measuring work-flow and its routineness, is likely to vary at minimum by level and department. Therefore, the sampling procedures should attempt to reflect these basic aspects of the organization.

Respondents within each organization were selected by the following criteria:

1. All department heads and supervisory personnel, both professional and non-professionl.

2. One half of other professionals in departments of less than ten members were selected randomly.

3. One third of other professionals in departments of more than ten members were selected randomly.

Non-supervisory administrative and maintenance personnel were not interviewed because they were less likely to be involved in the establishment of organizational goals and policies. The different ratios in larger and smaller departments insured that the smaller departments would be adequately represented. All supervisory personnel at the upper levels were included because they were most likely to be key decision-makers. The stratified design resulted in 11 interviews in the smallest organizations and 62 in one of the larger.

Since the unit of analysis was the organization, the data for each organization had to be aggregated in order to calculate scores for measures of organizational structures. This was done by first computing means for each social position in the organization. A social position was defined in the same way as the sample was stratified, that is, by level and department. For example, if an organization's professional staff consisted of two departments—psychiatric and social work—and two levels—supervisors and caseworkers—then there were four social positions in this organization.

Computation of means for each social position has the advantage of avoiding the problem potentially created by the use of two sampling ratios. In effect, responses are standardized by organizational location—level and occupation—and then combined into the organizational score. Computation of means of social position also has the major theoretical advantage of focusing on the sociological perspective of organizational reality. An organization is perceived as a collection of social positions rather than an aggregate of individuals.

The second step was to compute an organizational score for each dimension by averaging the means of all social positions, as defined, in each organization. Equal weight was thus given to each social position.

The sampling procedure omitted non-supervisory and non-professional personnel, but since this was done in all organizations it represents a standardized procedure. The sampling and aggregation procedures prevent the top echelons from being underweighted, which would happen if the score was computed by equally weighing each interview in an organization.

MEASUREMENT OF ROUTINENESS

Several different questions were used in constructing the measure of the routineness of work. The first of these questions is as follows: "Would you describe your work as being very routine, somewhat routine, somewhat nonroutine, or very nonroutine?" This question, together with several others developed by Hall (1963), loaded together in a factor analysis of various aspects of organizational behavior seem to represent a separate and distinct dimension representing the routineness of work. Hall's questions were:

1. People here do the same job in the same way every day.
2. One thing people like around here is the variety of work (reversed).
3. Most jobs have something new happening every day (reversed)
4. There is something different to do every day (reversed).

Respondents replied definitely true, true, false, or definitely false to each of the questions. A factor analysis was felt to be necessary because there is a seemingly close similarity between routineness of work process and other aspects of organizational life such as the degree of job codification or specification of work procedures. This variable has different substantive content from either of the measures of formalization used.

Routineness of work measures how much variety there is in work; job codification measures how well defined the job is; rule observation measures the enforcement of the rules; job specificity measures how concrete the job description or procedural manual is. Clearly, routineness is different from the content of each of these. As a check on this measure the distribution by type of organization was examined.

The organizations included here are relatively homogeneous since most of them provide psychological, psychiatric, or rehabilitation services of one kind or another. As one might expect in such organizations, the scores tend more towards nonroutineness. That is, the organizational scores ranged from 1.31 to 2.46 on a scale that could vary from 1.00 to 4.00.

The highest on routineness is a family agency in which the caseworkers use a standard client interview that takes less than fifteen minutes. The purpose of the interview is to ascertain the eligibility of clients for county, federal or state medical aid. One interviewee said: ". . . somewhat routine —Even though each patient is individual, the type of thing you do with them is the same. . . ." The organization at the other extreme is an elite psychiatric family agency in which each member is an experienced therapist and allowed to work with no supervision at all.

Information from other sources about the agencies within each organizational category reported in Table 3–1 suggests that the reports of the staff are relatively correct. For example, the residential treatment center that scored lowest on routineness is one in which the executive director reported that the work is so novel and innovative that his staff does not know which professional association to join because each of the existing ones is clearly inappropriate. One staff member of this organization replied: "Highly nonroutine. No two days are alike when you are working with highly unpredictable children. We individualize very highly here; we would turn the place upside down for one child—and sometimes we do." This organization was considering creating a new professional society. While there is little difference between degree of routineness among the three mental hospitals, the one that is least routine is an elite, private sanitorium attached to a university medical school.

In general, the range within each category of organizations is somewhat restricted except for the social casework agencies. This suggests that the sample is basically similar in terms of technology: each organization

TABLE 3-1
Means and Range for Scores of Degree of Routineness in Different Organizational Categories

Kind of organization	*Number*	*Mean*	*Range*
Family agencies	6	1.83	1.31-2.46
Rehabilitation centers	3	1.80	1.59-1.94
Residential treatment homes	3	1.73	1.46-1.90
Mental hospitals	3	1.73	1.63-1.82
Department of special education	1	1.64	1.64

relies heavily upon some variety of psychiatric techniques. The range is greatest in the social casework agencies because it is here that the standardized interview is most likely to be found. Examination of the responses of various occupational groups in the study gives greater appreciation of the validity of the measure and of the variation in technologies. As might be expected, the higher the level within each organization, the more likely the job occupant is to report that his job is very nonroutine (Hall, 1962). A supervisor in a residential treatment center said ". . . highly nonroutine because we're dealing with a rather flexible entity, that is, a disturbed student." A case coordinator in an elite social casework agency described his job as nonroutine because: ". . . the planning requires some creativity and supervision and consultation requires the ability to relate current problems which are never routine." Finally, a supervisor in a rehabilitation center replied: "Because of the kind of problems we have, they cannot be anticipated. Each week problems occur, so that appointments have to be changed constantly." But at each level there are differences. The supervisors of the ancillary departments, such as business or maintenance, are more likely to state that their jobs are routine than those who head departments of social work, psychology, nursing, or teaching. One accountant said: "By the nature of accounting and statistics—it must be routine—it's mandatory in accounting." This not only suggests that the sampling procedure is indeed an appropriate one, but that staff members more removed from dealing with the clients report more routine work procedures.

Members of occupational groups that spend a lot of time with clients were more likely to perceive their work as nonroutine than those who spend short periods of time with clients. Thus nurses are more likely than either psychiatrists or psychologists to state that their work is very nonroutine. Similarly, house parents who actually live with emotionally disturbed children are more likely to state nonroutine work than homemakers who visit homes for short periods of time.

Members of occupational groups that perform the same activities with clients without variation are more likely to perceive their work as routine than those who have greater variety in their work. For example, vocational instructors who teach standardized procedures were more likely to report their work as routine than teachers who teach verbal and written skills. One teacher in a residential treatment center said her work was nonroutine because of "working with emotionally disturbed children; from one day to the next their reaction varies." In contrast, a vocational instructor said her work was highly routine because: "You have to have a routine or you couldn't do it. You have to make a schedule and not interfere with anyone and you have to train your girls that way." Psychometricians were more likely to describe their work as routine than clinical psychologists. And, finally, administrative assistants were more apt to report routine work situations than assistant executive directors, even though they are on the same level of authority. The latter must supervise a variety of occupational skills.

While these examples demonstrate variations among different occupational groups and among different levels of authority in these organizations, the results of this study are not simply functions of variations in occupation and level of authority. Rather, the measure seems to have captured a characteristic of the organization itself. This contention is supported by our findings using one-way analysis of variance tests of three different independent variables: levels of authority (three), occupational groups (thirteen largest), and organization (the sixteen included here). Differences by level and occupational group were not statistically significant, but differences by organization were.

Hypotheses and Findings

ROUTINE WORK AND SOCIAL STRUCTURE

Perrow (1967:198–199) begins his discussions of the impact of technology on the social structure (he uses the phrase "structure task") by noting that coordination can occur either via planning, programmed interaction, or feedback. This basic assumption, which is derived from March and Simon (1958: chapter 6) provides a basis for interpreting the possible consequences of the degree of routineness of work. If technology can be routinized, then coordination can be and probably will be planned and programmed. If it cannot, then coordination must be effected via feedback. From this flows a series of consequences for organizational social structure.

1. *Organizations with routine work are more likely to be characterized by centralization of organizational power.*

If organizational members constantly face a work situation characterized by highly varied clients' need, then greater organizational power will accrue to organizational members who interact with the clients most frequently. It will be necessary to have constant reports from such individuals in order for adequate coordination to be accomplished. The need for feedback about client-handling and particularly about basic policies, as well as the development of new programs, will result in a decentralization of organizational decision-making (cf. Greenblatt, York and Brown, 1955). The search for new programs is presumedly a reflection of the attempt to find more adequate technology for solving the problems of the clients. But this requires a series of decisions about personnel, policies, programs, and budget allocations. The most appropriate participants in making such decisions are those who work with clients.

The findings shown in Table 3–2 suggest that this line of reasoning is correct. The correlation coefficient between the degree of routineness of work and participation in organizational decision-making is —.72; that is, the more routine the work-flow, the greater the centralization of decision-making about basic organizational issues.

It should be noted that in his original discussion, Perrow suggested four possible arrangements of an organization's power structure, corresponding to his two basic continua defining routineness: the incidence of exceptional cases and the incidence of analyzable problems. In the extreme case of nonroutine work, that is, where there is both variability in clients as well as lack of knowledge about their handling, the power structure

TABLE 3-2
The Relationship between Routine Work and Social Structure*

Structural variables	*Pearsonian product-moment correlation coefficients*	*p*
Degree of centralization		
Degree of participation in organizational decisions	–.72	<.001
Degree of hierarchy of authority in work decisions	–.02	N.S.
Degree of formalization		
Degree of job codification	.21	N.S.
Degree of rule observation	.20	N.S.
Presence of a rules manual	.51	<.05
Presence of job descriptions	.53	<.05
Degree of specificity of job descriptions	.61	<.01
Degree of stratification		
Affect between supervisors and staff	–.09	N.S.
Distance between supervisors and staff	.19	N.S.
Degree of complexity		
Amount of professional training	–.55	<.05
Amount of professional activity	–.12	N.S.
Number of occupational specialties	–.19	N.S.

*For discussion of the construction of all the measures of the variables included here with the exception of the presence of a rules manual and job descriptions and the two measures of stratification, see Aiken and Hage, 1968.

should be polycentralized (Perrow, 1967:Figure 3). The data in Table 3–2 indicate that it is decentralized. (The polycentralized organization is more probable in Perrow's type 1 organization.) Actually, Perrow's own discussion of the high power of both technicians and supervisors indicates that the nonroutine organization really should be decentralized since he states that both groups will have high power.

Perrow (1967:198) makes a critical distinction between organizational power and work discretion. This corresponds to discussions of other researchers, such as Blauner (1964), who have noted the differences between participation in basic organizational decisions about policies and participation in decisions about work appropriate to a particular job (cf. Hage and Aiken, 1967). Perrow indicates that routine work might lead both to low discretion over work decisions as well as little power in making organizational decisions. The measure of autonomy in work decisions used in this study—the hierarchy of authority—indicates that this is not necessarily the case. There was no correlation ($r=-.02$) between these two organizational characteristics. Discretion over work is evidently affected by other organizational variables, not by the degree of routineness of work.

2. *Organizations with routine work are more likely to have greater formalization of organizational roles.*

If an organization is coordinated via planning, this will be reflected in a high degree of formalization of organizational roles. Such formalization will be reflected in various types of rules and regulations defining role obligations and enactment. One way to bring about such formalization is to establish a series of written documents that specify who is to do what, when, where, and why. Routine work obviously facilitates the process of formalization by providing the stability or lack of variety in clients that makes the writing of documents containing regulations or rules more manageable. Such documents are variously called policy manuals, job descriptions or evaluation procedures.

The use of a rules manual and of job descriptions are two basic mechanisms for establishing organizational control. Aside from these indicators of formalization, the degree of specificity in job descriptions, the degree of codification of jobs, and the degree of rule observation, that is, the degree of surveillance, were measured. The major distinction between job codification and specificity of the job descriptions is one between rules and procedures; the former measures whether or not there are rules for the job occupants while the latter measures the degree to which work procedures are specified. These are complementary ideas that need not necessarily be correlated (Rushing, 1966). Indeed, each of these three dimensions was obtained in a factor analysis of several batteries of tests developed by Hall (1963). Since there is such a close connection between the idea of routineness and formalization, items in these dimensions were included in the same factor analysis, and these items loaded with separate factors. The presence of a rules manual and of job descriptions was based on staff responses to questions about each of these. The items included in the other three variables are described elsewhere (Aiken and Hage, 1968).

Routine work leads to the formalization of regulations as represented by the presence of a rules manual (r = .51), the presence of job descriptions (r = .53), and the degree of job specificity (r = .61), but has little impact on job codifications (r = .21) and rule observation (r = .20) although the relationships with these factors are in the predicted direction (see Table 3–2). This parallels the findings on the relationship between routine work, power, and discretion over work. Job codification and rule observation refer more to work regulations while the rules manual, job descriptions and job specificity reflect basic work guidelines and policies of the organization. These last three are more likely to be manifestations of the programming for the organization's coordination than the other two measures.

3. *There is no relationship between the degree of routine work and organizational stratification.*

If the organization is highly programmed with routine work, there is little need for interaction; the interaction that does occur can be programmed in ritualized reports to the boss. Perrow (1967: Figure 3) hypothesizes that there is less interdependence of work groups in a routine organization. This suggests that there may be considerable social distance between levels of hierarchy or chain command. An excellent illustration of this phenomenon is found in Crozier's (1964) study of bureaucracy where there was considerable difference in the prestige of staff and supervisors in an obviously routine organization.

Two dimensions of organizational stratification, the degree of affect toward supervisors and the degree of social distance between supervisors and staff were measured. These dimensions were obtained from a factor analysis of scales developed by Seeman and Evans (1961) in their study of the stratification of medical wards. The findings indicate that there is no support for this hypothesis. There is almost no relation between the degree of affect and routineness (r = .09), and there is only a very weak positive relationship between routineness and social distance (r = .19), indicating that the direction of these relationships support the expectation of barriers developing between supervisors and staff when the work-flow is routinized, but that the strength of these relationships are not large enough to argue for support of the hypothesis.

4. *Organizations with routine work are likely to have staff with less professional training.*

Although Perrow (1967) does not discuss this dimension of organizational structure, it would appear that if the organization is coordinated by programming, there is less need for a variety of occupational specialties or for well-trained occupants. In other words, the programming of the organization allows for simplification of the social structure. As technology becomes routinized, it is less neccessary for an organization to hire men with highly specialized skills. The main purpose of the assembly-line, the extreme routinization of work-flow, is to eliminate completely the need for highly skilled labor. To a lesser extent, the same principle can operate in people-processing organizations. If the interview of a client is straightforward, for example, simply ascertaining the eligibility of the client for welfare, then the organization has little need for job occupants with advanced degrees. In contrast, using psychiatric skills requires at a minimum a master's degree, and more advanced training is often desirable. To a lesser extent, the routinization of technology can affect the number of different occupational specialties as well. If little is known about the particular technology, the organization is likely to hire different specialties in the hope that something may work. As one member of a rehabilitation organization said, "We keep thinking that if we have a variety of specialists, our clients are better off, but in fact we don't know if this is really true."

Organizations with routine work have less well-trained staff. The correlation between professional training and degree of routineness is —.55. But the degree of routine work has little affect on either the diversity of occupational groups in an organization ($r = -.19$) or on the relationship with the amount of professional activity, the third measure of complexity ($r = -.12$). Routine work reduces the need for expertise but does not seem to affect either the variety of job specialties or the amount of involvement in professional activities.

The routineness of work may be only a consequence of size. That is, as size of the organization increases bureaucratization might also increase, thus providing a strong pressure toward routinization of work. This would mean that many of the observed relationships between this aspect of technology and these social structural variables may simply be functions of the inevitable programming that is necessary for the coordination of larger organizations. Almost no relationship was found between the size of the organization and the routineness of the work process (r = .07). This finding indicates that relationships between the routineness of work and other structural properties of organizations are evidently not a function of the size of the organization. Routineness of technology should be treated as an input that can affect the social structure of an organization independently of organizational size (Pugh *et al.*, 1969).

ROUTINE WORK AND ORGANIZATIONAL GOALS

Earlier it was suggested that technology can set limits on the goals of the organization. This process presumedly occurs via the structure of the organization. As technology is routinized, the organization is coordinated via programming. This is likely to be accompanied by centralization of power, formalization of roles, and some lessening of the level of professionalization in the organization. In such a social structure, a limited range of goals are likely to be maximized (Hage, 1965). In such a structure the organization is likely to emphasize the goal of efficiency in preference to other goals such as morale. Similarly, organizations with routine work structures are likely to emphasize the quantity of clients serviced rather than the quality of those services.

Perrow (1967:202) suggests basically the same hypotheses, although he does not provide any rationale for them. He does make a crucial distinction between system goals and product goals. In Perrow's view, the routine organization is likely to be concerned with stability and high profits achieved via quantity of production and an avoidance of innovation. In contrast, the nonroutine organization will emphasize growth, quality, and innovation, being less concerned with making profits. While these are examples of industrial firms, they suggest consequences for people-processing organizations as well, since the referents of system and product goals can be applied to all kinds of organizations, the idea is one of service. Do the organizational decision-makers emphasize the quantity or the quality of client service? Similarly, the emphasis on efficiency is akin to the idea of profits while emphasis on new programs is akin to the idea of innovation. Thus, while the system goals included here are not exactly the same as Perrow's, they are similar in meaning and can be applied to both industrial and non-industrial organizations.

This discussion of goals, whether system or product goals, is more general than the usual discussion of goals where the referent is the specific goal of some type of organization (cf. March, 1965; Zald and Denton, 1963). Thus, the goal of schools is to teach or socialize; the goal of hospitals is to cure; the goal of business firms is to make profits, and so on.

Similarly, the measures used in this study avoid this parochial approach to the study of goals by listing basic dimensions that can be applied to any organization, whether profit-making or not, thus providing an opportunity for the development of general propositions across all kinds of organization.

Respondents in the organizations were asked in a series of six paired comparisons to select the goal their organization was emphasizing at the time of interview: (1) the effectiveness of client services, (2) the efficiency of operation, (3) the morale of the staff, or (4) the development of new programs or services.

To minimize the effect of response set, the ordering of presentation of pairs was randomized. It should be noted that these questions are designed to elicit reports about the organization's emphasis on goals. In order to measure product goals as described by Perrow, the respondents were additionally asked whether the organization placed greater emphasis on the quantity of clients served or on the quality of client service. There is the possibility that the respondents reported the organizational priorities espoused by the organizational elites rather than reporting the actual emphasis in the organization for their point of view. That is, there is the danger that they simply reported the goals desired by organizational leaders. There is no assurance that actual outputs were measured. Values about goals is a safer term to describe what was measured.

A scale was constructed for each goal such that in each of the three comparisons made, ten points were assigned if the respondent selected the goal and five points if he said half and half or gave both goals equal weight. Thus the scale for each goal varies from zero (the goal was never chosen) to thirty (the goal was chosen in each of the three pairs). A separate scale was constructed for each of the four goals.

There is no association between routineness and relative effectiveness as an organizational goal ($r = .00$). This is not unexpected since the label "effectiveness" probably elicited normative responses. Most organizations strive to maximize effectiveness, but they differ in the pathways they choose; some choose to emphasize morale, others efficiency and still others new programs.

1. *Routine work is positively related to an emphasis on efficiency as a system goal.*

The routinization of the work-flow allows for the efficient handling of large numbers of clients at low cost, efficiency thereby becoming a systems goal. The notion of efficiency in nonprofit organizations is analogous to that of the profit motive in a business. Efficiency is facilitated via the formalization of regulations and careful planning of the organization. Table 3–3 indicates that these two organizational variables—degree of routineness and emphasis on efficiency—are related (r = .45).

2. *Routine work is negatively, but weakly related to morale as a system goal.*

The great strength of a routine work-flow is that it allows for the careful coordination of all the organizational tasks. But in the process, the fact that human beings are involved is frequently neglected. Thus routine work results in a lessening of emphasis on morale as a system goal. In addition, in many circumstances there is likely to be an incompatibility between the goals of morale and efficiency. Fringe benefits, higher salaries, and favorable working conditions may not be consistent with efficiency, but they affect morale. Perhaps more importantly, the pressure toward rapid, low cost production (that is, efficiency) reduces morale. Similarly, regulations which are the *sine qua non* of efficiency debilitate morale and enmesh staff members in red tape.

As expected, there is an inverse relationship between the degree of routine work and the emphasis on morale (r = −.37), but this relationship is not statistically significant. Organizations in which there is a routine work-flow are more likely to be those in which policy does not maximize high morale, at least according to the staff members in those organizations, but this is only a tendency since this relationship is not strong.

3. *There is no relationship between routine work and the development of new programs as a system goal.*

Much of the reasoning discussed above applies to this goal. New programs do not allow for an emphasis on efficiency. Similarly, routine work is the antithesis of the nature of work in the innovating organization (Wilson, 1966).

TABLE 3-3
The Relationship between Routine Work and Organizational Goals

System goals	*r*	*p*
Emphasis on effectiveness	.00	N.S.
Emphasis on efficiency	.45	.10
Emphasis on morale	−.37	N.S.
Emphasis on new programs	−.05	N.S.
Product-characteristic goals		
Emphasis on quality	−.42	N.S.

Despite the plausibility of the argument that the organization that is emphasizing new programs is one interested in innovativeness, the data do not support the hypothesis. There is no correlation between the emphasis on the development of new programs and the routineness of work ($r = -.05$). Admittedly, the development of new programs is not the only way in which organizations can innovate.

4. *Routine work is negatively, but weakly related to the quality of client service as a product goal.*

A routine work-flow allows for the gradual speed-up of the work process. The difficulty with the assembly-line and its speed of operation is the clearest example of this phenomenon, but the Blau (1965) study of records in a welfare agency illustrates the same idea in the context of people-processing organizations. The consequence of this is the maximization of the goal of quantity as opposed to quality.

The hypothesis is moderately supported by the data ($r = -.42$). The more that technology is routinized, the less the emphasis on the quality of service, and the more the emphasis on the number of clients served.

Summary and Discussion

Most of Perrow's implicit hypotheses about the relationships between the routineness of technology and dimensions of social structure receive considerable support. The more routine the organization, the more centralized the decision-making about organizational policies, the more likely the presence of a rules manual and job descriptions, and the more specified the job. While the measures of stratification used seem to measure a concept similar to Perrow's idea of interdependence, these concepts are distinct, so the lack of supporting findings here do not necessarily provide negative evidence. Although Perrow did not discuss the degree of complexity, it is interesting to note that routineness is negatively related to the amount of professional training but does not appear to have much association with other measures of complexity. Similarly, routineness does not appear to have much association with the hierarchy of authority or discretion over work decisions.

When the relationship between routineness of technology and organizational goals is examined, there are additional confirming results. In organizations with a relatively routine technology, the staff members are likely to report more emphasis on efficiency and quantity of clients than on quality of service and staff morale. There is no association between routineness of technology and the goal of innovation measured in this study.

It should be remembered that most of the 16 organizations studied were on the nonroutine side of the scale. But, precisely because it is a relatively homogeneous sample, the findings are all the more encouraging. The results do not appear to be simply a consequence of size; size has little or no relationship with routineness. While it is relatively easy to accept hypotheses about the relationship between routineness and properties of an organizational structure, it is less easy to accept the idea that the degree of routineness is the only variable involved. The size of the organization, its autonomy, and its level of financing are probably variables of equal importance.

Nor should the routineness of technology be considered as the only relevant technological dimension. Other dimensions such as the number of different kinds of technologies or the amount of organizational knowledge also need to be explored. In particular, it may be that these are more likely to be related to measures of complexity and the amount of organizational innovations.

Other input variables besides technology need to be explored in order to understand the relative importance of technology in determining social structure and in setting limits on organizational policy.

REFERENCES

Aiken, Michael, and Jerald Hage
1968 "Organizational interdependence and intra-organizational structure." American Sociological Review, 33 (December): 912–930.
Blau, Peter
1955 The Dynamics of Bureaucracy. Chicago: University of Chicago Press.
Blau, Peter, and W. Richard Scott
1962 Formal Organizations. San Francisco: Chandler.
Blauner, Robert
1964 Alienation and Freedom: The Factory Worker and His Industry. Chicago: University of Chicago Press.
Coleman, James S.
1964 "Research chronicle: the adolescent society." In Phillip Hammond (ed.), Sociologist at Work: 213–243. New York: Basic Books.
Crozier, Michel
1964 The Bureaucratic Phenomenon. Chicago: University of Chicago Press.
Greenblatt, Milton, Richard York, and Esther Brown
1955 From Custodial to Therapeutic Patient Care in Mental Hospitals. New York: Russell Sage Foundation.
Hage, Jerald
1965 "An axiomatic theory of organizations." Administrative Science Quarterly, 10 (December): 289–321.

Hage, Jerald, and Michael Aiken
1967 "Relationship of centralization to other structural properties." Administrative Science Quarterly, 12 (June): 72–92.
Hall, Richard
1962 "Intraorganizational structural variation: application of the bureaucratic model." Administrative Science Quarterly, 7 (December): 295–308.
1963 "The concept of bureaucracy: an empirical assessment." American Journal of Sociology, 69 (July): 32–40.
Labovitz, Sanford
1963 "Technology and Division of Labor." Doctoral dissertation, University of Texas.
Lazarfeld, Paul, and Herbert Menzel
1961 "On the relation between individual and collective properties." In Amitai Etzioni (ed.), Complex Organizations: A Sociological Reader: 422–440. New York: Holt, Rinehart, and Winston.
Litwak, Eugene
1961 "Models of bureaucracy which permit conflict." American Journal of Sociology, 67 (September): 177–184.
March, James G.
1965 The Handbook of Organizations. Chicago: Rand McNally.
March, James G., and Herbert A. Simon
1958 Organizations. New York: Wiley.
Parsons, Talcott, and Neil J. Smelser
1956 Economy and Society: A Study in the Integration of Economic and Social Theory. New York: Free Press.
Perrow, Charles
1967 "A framework for the comparative analysis of organizations." American Sociological Review, 32 (April): 194–208.
Pugh, Derek, David J. Hickson, and C. Robin Hinings
1969 "The context of organization structures." Administrative Science Quarterly, 14 (March): 91–114.
Rushing, William
1966 "Organizational rules and surveillance." Administrative Science Quarterly, 10 (March): 423–443.
Seeman, Melvin, and John Evans
1961 "Stratification and hospital care: I. The performance of the medical interne." American Sociological Review, 26 (February): 67–80.
Thompson, Victor
1961 Modern Organization. New York: Knopf.
Wilson, James Q.
1966 "Innovation in organization: notes toward a theory." In James Thompson (ed.), Approaches to Organizational Design: 193–218. Pittsburgh: University of Pittsburgh Press.
Woodward, Joan
1965 Industrial Organization. London: Oxford University Press.
Zald, Mayer
1963 "Comparative analysis and measurement of organizational goals: the case of correctional institutions for delinquents." Sociological Quarterly, 4 (Summer): 206–230.
Zald, Mayer, and Patricia Denton
1963 "From evangelism to general service, the transformation of the YMCA." Administrative Science Quarterly, 8 (September): 214–234.

4

D. S. PUGH, D. J. HICKSON,
C. R. HININGS, AND C. TURNER

THE CONTEXT OF ORGANIZATION STRUCTURES

The structure of an organization is closely related to the context within which it functions, and much of the variation in organization structures might be explained by contextual factors. Many such factors, including size, technology, organizational charter or social function, and interdependence with other organizations, have been suggested as being of primary importance in influencing the structure and functioning of an organization.

There have been few attempts, however, to relate these factors in a comparative systematic way to the characteristic aspects of structure, for such studies would require a multivariate factorial approach in both context and structure. The limitations of a unitary approach to organizational structure have been elaborated elsewhere (Hinings *et al.*, 1967), but its deficiencies in the study of contextual factors are no less clear. Theorists in this area seem to have proceeded on the assumption that one particular contextual feature is the major determinant of structure, with the implication that they considered the other less important. Many writers from Weber onwards have mentioned size as being one of the most important causes of differences between structures, and large size has even been considered as characteristic of bureaucratic structure (Presthus, 1958). Others argue for the pre-emptive importance of the technology of production or service in determining structure and functioning (Dubin, 1958; Perrow, 1967; Woodward, 1965; Trist *et al.*, 1963). Parsons (1956) and Selznick

Reprinted from D. S. Pugh, D. J. Hickson, C. R. Hinings, and C. Turner, "The Context of Organization Structures," *Administrative Science Quarterly*, 14–1 (March, 1969), pp. 91–113.

(1949) have attempted to show in some detail that the structure and functioning of the organization follow from its social function, goals, or "charter." Eisenstadt (1959) emphasized the importance of the dependence of the organization on its social setting, particularly its dependence on external resources and power, in influencing structural characteristics and activities. Clearly all of these contextual factors, as well as others, are relevant; but without a multivariate approach, it is not possible to assess their relative importance.

A previous paper described the conceptual framework upon which the present multivariate analysis is based (Pugh *et al.*, 1963), and a subsequent paper its empirical development (Pugh *et al.*, 1968). It is not a model of organization in an environment, but a separation of variables of structure and of organizational performance from other variables commonly hypothesized to be related to them, which are called "contextual"

TABLE 4-1
Conceptual Scheme for Empirical Study of Work Organizations

Contextual variables	*Structural variables*†
Origin and history	*Structuring of activities*
Ownership and control	Functional specialization
Size	Role specialization
Charter	Standardization (overall)
Technology	Formalization (overall)
Location	*Concentration of authority*
Resources	Centralization of decision making
Dependence	Autonomy of the organization
*Activity variables**	Standardization of procedures for selection and advancement
Identification (charter, image)	*Line control of workflow*
Perpetuation (thoughtways, finance, personnel services)	Subordinate ratio
Workflow (production, distribution)	Formalization of role performance recording
Control (direction, motivation, evaluation, communication)	Percentage of workflow superordinates
Hemeostatis (fusion, leadership, problem solving, legitimization)	*Relative size of supportive component*
	Percentage of clerks
	Percentage of non-workflow personnel
	Vertical span (height)
	Performance variables
	Efficiency (profitability, productivity, market standing)
	Adaptability
	Morale

*Bakke (1959).
†Pugh *et al.* (1968).

in the sense that they can be regarded as a setting within which structure is developed. Table 4–1 summarizes the framework and also includes a classification of activities useful in the analysis of organization functioning (Bakke, 1959).

The design of the study reported in the present paper treats the contextual variables as independent and the structural variables as dependent. The structural variables are (1) *structuring of activities*; that is, the degree to which the intended behavior of employees is overtly defined by task specialization, standard routines, and formal paper work; (2) *concentration of authority*; that is, the degree to which authority for decisions rests in controlling units outside the organization and is centralized at the higher hierarchical levels within it; and (3) *line control of workflow*; that is, the degree to which control is exercised by line personnel instead of through impersonal procedures. The eight contextual variables were translated into operational definitions and scales were constructed for each of them. These were then used in a multivariate regression analysis to predict the structural dimensions found.

This factorial study using cross-sectional data does not in itself test hypotheses about *processes* (e.g. how changes in size interact with variations in structuring of activities), but it affords a basis for generating such hypotheses.

Sample and Methods

Data were collected on fifty-two work organizations, forty-six of which were a random sample stratified by size and product or purpose. The sample and methods have been described in detail in a previous paper, (Pugh *et al.*, 1968). For scaling purposes, data on the whole group were used, but for correlational analyses relating scales to each other, and for prediction analyses relating contextual variables to structural ones, only data on the sample of forty-six organizations were used. None of the data was attitudinal.

The data were analyzed under the heading of the conceptual scheme. To define the variables operationally, scales were constructed that measured the degree of a particular characteristic. The scales varied widely. Inkson *et al.* (1967) discussed the variety of scaling procedures used. Some were simple dichotomies (such as impersonality of origin) or counts (such as number of operating sites); some were ordered category scales, locating an organization at one point along a postulated dimension (such as closeness of link with customers of clients). Some were stable, ordered scales established by linking together a large number or items exhibiting the characteristic on the basis of cumulative scaling procedures, such as workflow rigidity, an aspect of technology. Some were summary scales extracted by principal-components analysis to summarize a whole dimension, such as operating variability, an aspect of charter. In this way, forty primary scales of context were constructed and then reduced to fourteen empirically distinct elements, which are listed in Table 4–2 together with their correlations with the main structural variables as defined in Table 4–1. Table 4–3 gives their intercorrelations. The methodological implications of this analysis are discussed in Levy and Pugh (1969).

The study of contextual aspects of organizations will inevitably produce a much more heterogeneous set of scales than the comparable study of the

TABLE 4-2
Elements of Organization Context

	Product-moment correlation with structural factors		
Elements of context	*Structuring of activities*	*Concentration of authority*	*Line control of workflow*
*Origin and history (3)**			
Impersonality of origin	–0.04	0.64	0.36
Age	0.09	–0.38	–0.02
Historical changes	0.17	–0.45	–0.03
Ownership and control (7)			
Public accountability	–0.10	0.64	0.47
Concentration of ownership with control	–0.15	–0.29	–0.21
Size (3)			
Size of organization†	0.69	–0.10	–0.15
Size of parent organization†	0.39	0.39	–0.07
Charter (7)			
Operating variability	0.15	–0.22	–0.57
Operating diversity	0.26	–0.30	–0.04
Technology (6)			
Workflow integration	0.34	–0.30	–0.46
Labor costs	–0.25	0.43	0.32
Location (1)			
Number of operating sites	–0.26	0.39	0.39
Dependence (10)			
Dependence	–0.05	0.66	0.13
Recognition of trade unions	0.51	0.08	–0.35

*Numbers in parentheses indicate number of primary scales.
†Logarithm of number of employees.

TABLE 4-3
Intercorrelations of Contextual Variables (Product-moment Coefficients, $N = 46$)

Scale title	Impersonality of origin	Age	Historical changes	Public accountability	Concentration of Ownership with control*	Size of organization†	Size of parent organization†	Operating variability	Operating diversity	Workflow integration	Labor costs	No. of operating sites	Dependence	Trade unions
Impersonality of origin	—													
Age	−0.20	—												
Historical changes	−0.34	0.50	—											
Public accountability	0.66	0.00	−0.25	—										
Concentration of ownership with control*	−0.40	−0.03	0.02	−0.50	—									
Size of organization†	0.07	0.16	0.29	0.00	−0.21	—								
Size of parent organization†	0.45	−0.12	−0.10	0.51	−0.55	0.43	—							
Operating variability	−0.26	−0.24	−0.16	−0.34	0.29	−0.24	−0.19	—						
Operating diversity	−0.23	0.00	0.13	−0.14	0.00	0.26	−0.10	0.02	—					
Workflow integration	−0.24	0.07	0.05	−0.35	0.10	0.07	−0.09	0.57	0.33	—				
Labor costs	0.41	−0.24	−0.31	0.34	−0.09	−0.28	0.08	−0.27	0.01	−0.50	—			
No. of operating sites	0.14	−0.07	−0.08	0.34	−0.20	0.14	0.16	−0.56	−0.05	−0.58	0.16	—		
Dependence	0.53	−0.32	−0.38	0.53	−0.50	−0.17	0.63	0.05	−0.19	−0.05	0.26	0.05	—	
Recognition of trade unions	0.04	−0.04	−0.11	0.17	−0.21	0.36	0.37	0.19	0.01	0.20	−0.15	−0.12	0.22	—

*$N = 42$ for all correlations with this variable.

†Logarithm of number of employees.

structural aspects; for the scales are selected, not from a common conceptual base, but for their postulated links with structure. One of the objectives of using the multivariate approach described here would be to test the relationship between disparate aspects of context, and to attempt a conceptual clarification of those aspects demonstrated to be salient in relation to organizational structure.

It was not possible to investigate the variable "resources" adequately. For human and ideational resources, the wide-ranging interviews within a comparatively short time span made it impossible to obtain adequate data. Material and capital resources were found to reduce to aspects of size, and the relative disposition of these resources (e.g. capital versus labor) was found to be better regarded as an aspect of technology.

Contextual Variables

ORIGIN AND HISTORY

An organization may have grown from a one-man business over a long period of time, or it may have been set up as a branch of an already existing organization and so develop rapidly. During its development it may have undergone many or few radical changes in purpose, ownership, and other contextual aspects. An adequate study of the impact of these factors on organizational structure must be conducted on a comparative longitudinal basis (Chandler, 1962); but even in a cross-sectional study such as this, it is possible to define and make operational three aspects of this concept.

Impersonality of Origin. This variable distinguishes between entrepreneurial organizations, personally founded, and bureaucratic ones founded by an existing organization. Impersonally founded organizations might be expected to have a higher level of structuring of activities, whereas personally founded organizations would have a higher degree of concentration of authority. The data on the present sample, however, show no relationship between impersonality of origin and structuring of activities ($r = -0.04$), but a strong relationship between impersonality of origin and concentration of authority ($r = 0.64$). (With $N = 46$, all correlations 0.29 and above are at or beyond the ninety-five percent level of confidence.) To a considerable extent this relationship is due to the fact that government-owned, and therefore impersonally founded, organizations tend to be highly centralized. Such organizations tend to be line controlled in their workflow, thus contributing to the relationship

($r = 0.36$) between impersonality of origin and line control of workflow. The lack of relationship with structuring of activities, which is common to all three scales of this dimension, underlines the need to examine present contextual aspects in relation to this factor rather than historical ones.

Age. The age of the organization was taken from the time at which the field work was carried out. The range in the sample varied from an established metal goods manufacturing organization, founded over 170 years previously, in 1794, to a government inspection department, which began activities in the area as a separate operating unit 29 years previously. No clear relationship was found between age and impersonality of origin ($r = -0.20$). Stinchcombe (1965) has argued that no relationship should be expected between the age of an organization and its structure but rather between the structure of an organization and the date that its industry was founded. The present data support this conclusion in that no relationship is found between age and structuring of activities ($r = 0.09$) or line control of workflow ($r = 0.02$). Age was related to concentration of authority ($r = -0.38$), older organizations having a tendency to be more decentralized and to have more autonomy.

Historical Changes. The organizations in this sample did not have adequate historical information on the extent of contextual changes for use in a cross-sectional investigation; but it was possible to obtain limited information as to whether particular changes had occurred, and thus to develop a scale for the *types* of contextual changes that had occurred, namely whether at least one change had occurred (1) in the location of the organization, (2) in the product or service range offered, and (3) in the pattern of ownership. Item analysis carried out using the Brogden-Clemans coefficient (Brogden, 1949) gave a mean item-analysis value of 0.85, suggesting that it was possible to produce a scale of historical changes by summing the items. The organizations were distributed along the scale from no changes to all three types of changes. As expected, there was a strong correlation of this scale with age ($r = 0.51$), older organizations tending to have experienced more types of change. There was also a strong relationship, perhaps mediated by age, between historical changes and concentration of authority ($r = -0.45$), such changes being associated with dispersion of authority.

OWNERSHIP AND CONTROL

The differences in structure between a department of the government and a private business will be due to some extent to the different ownership and control patterns. Two aspects of this concept, public accountability and the relationship of the ownership to the management of the organization were investigated. For wholly owned subsidiary companies, branch factories, local government departments, etc. this form of analysis had to be applied to the parent institution exercising owning rights, in some cases through more than one intermediate institution (e.g. committees of the corporation, area boards, parent operating companies, which were themselves owned by holding companies, etc.). The ultimate owning unit is referred to as the "parent organization."

Public Accountability. This was a three-point category scale concerned with the degree to which the parent organization, (which could, of course, be the organizational unit itself, as it was in eight cases) was subject to public scrutiny in the conduct of its affairs. Least publicly accountable would be a company not quoted on the stock exchange; next, organizations that raised money publicly by having equity capital quoted on the stock exchange, also public cooperative societies; and most publicly accountable were the departments of the local and central government. On the basis of the classical literature on bureaucracy as a societal phenomenon, it might be hypothesized that organizations with the greatest exposure to public accountability would have a higher degree of structuring of activities, and a greater concentration of authority. The data on the present sample show relationships more complicated than this, however.

First, it must be emphasized that although this sample included eight government departments, all the organizations had a nonadministrative purpose, which could be identified as a workflow (Pugh, *et al.*, 1968: Table 1). This is not surprising in this provincial sample, since purely administrative units of the requisite size (i.e., employing more than 250 people) are few outside the capital. The relationships between public accountability and structure must be interpreted in the light of this particular sample.

No relationship was found between public accountability and structuring of activities ($r = -0.10$). This structuring factor applies to the workflow as well as administrative activities of the organization, and it appears that government organizations with a workflow are not differentiated from nongovernment organizations on this basis. On the other hand there was a positive relationship between public accountability and concentration of authority ($r=0.63$), standardization of procedures for selection and advancement ($r=0.56$), and line control of workflow ($r=0.47$). These all point to centralized but line-controlled government workflow organizations (Pugh *et al.*, 1969). The scale of standardization was a

›ipolar one, and a high score meant that the organization standardized its ›rocedures for personnel selection and advancement, and also that it did *ıot* standardize its procedures for workflow. The relationship between ›ublic accountability and this standardization scale suggests that the gov-ːrnment workflow organizations standardize their personnel procedures, ›ut rely on professional line superordinates for workflow control.

Relationship of Ownership to Management. The concepts of Sargent ᶠlorence (1961) were found most fruitful in studying this aspect of owner-hip and control, but the method used was the selection of variables for a :orrelational approach, rather than classification on the basis of percent-ːges. Florence studied the relationships of shareholders, directors, and ːxecutives. Where these groups were completely separate there was full ·eparation of ownership, control, and management; where they were the ame, then ownership, control, and management coalesced. Between these wo extremes, the scales were designed in the present study to measure he degree of separation. Company records and public records were exam-ned and five scales developed for the patterns of shareholding and the ·elationships between the ownership and the management of the organiza-ion. For the four foreign-owned organizations in the sample, this informa-ion was not available in England; the analysis was therefore based on ı = 42 (Table 4-4). A sixth scale was developed for interlocking ·irectorships; that is, the percentage of directors who held other director-hips outside the owning group. The intercorrelation matrix of these six ariables suggested that factor analysis would be helpful in summarizing n extensive analysis of ownership (Table 4–5). A principal-components nalysis was thus applied to the matrix, and a large first factor accounting or 56 percent of the variance was extracted, which was heavily loaded ·n all variables except interlocking directorships and was therefore termed concentration of ownership with control."

As would be expected, there was a negative relationship between public .ccountability and concentration of ownership with control ($r = 051$); he more publicly accountable the ownership, the less concentrated it ·as, with central and local government ownership epitomizing diffuse ·wnership by the voting public.

The discussion about the effects of differing patterns of personal owner-hip on organizations and society originated with Marx, and has since ·olarized into what Dahrendorf (1959) has called the "radical" and ·conservative" positions. It is generally agreed that there has been a rogressive dispersion of share ownership following the rise of the corpo-ation, but there is little agreement, or systematic evidence, on the effects

TABLE 4-4
Ownership and Control ($N = 42$)

	Scale number and title	*Range %*	*Mean*	*S.D.*
12.01	Concentration of voteholdings (Percentage of equity owned by top twenty shareholders)	0-100	38.47	32.37
12.03	Voteholdings of individuals (Percentage of individuals among top twenty shareholders)	0-100	17.19	26.89
12.04	Directors among top twenty voteholders (Percentage of directors among top twenty shareholders)	0-100	20.69	29.39
12.05	Directors' voteholdings (Percentage of equity owned by all directors combined)	0- 99.9	9.40	19.61
12.06	Percentage of directors who are executives	0-100	46.11	32.73
12.09	Interlocking directorships (Percentage of directors with other directorships beyond owning organization)	0-100	45.22	33.73

TABLE 4-5
Ownership and Control: Intercorrelation Matrix (Product-moment Coefficients, $N = 42$)

	Concentration of voteholdings	*Voteholdings of individuals*	*Directors among top twenty voteholders*	*Directors' voteholdings*	*Percentage of directors who are executives*	*Interlocking directorships*
Concentration of voteholdings	—					
Voteholdings of individuals	0.62	—				
Directors among top twenty voteholders	0.54	0.87	—			
Directors' voteholdings	0.55	0.90	0.78	—		
Percentage of directors who are executives	0.26	0.30	0.37	0.20	—	
Interlocking directorships	0.32	0.03	0.04	0.09	0.33	—

of this. The radicals (Burnham, 1962; Berle and Means, 1937) argue that present ownership patterns have produced a shift in control away from the entrepreneur to managers, who become important because of their control over the means of production and the organization of men, materials, and equipment. The result then of dispersion of ownership is likely to be dispersion of authority. However, the conservatives (Mills, 1956; Aaronovitch, 1961) argue that the dispersion of capital ownership makes possible the concentration of economic power in fewer hands, because of the inability of the mass of shareholders to act, resulting in a concentration of authority.

The results obtained with this sample support neither of these positions. The correlation given in Table 4–2 of concentration of ownership with control with concentration of authority ($r = -0.29$) might suggest that concentration of ownership is associated with dispersion of authority; but it must be remembered that this correlation is obtained for the whole sample, which includes government-owned organizations, whereas the discussion of the effects of ownership patterns has been concerned entirely with private ownership. When the government organizations were extracted from the sample, the correlation disappeared ($r = -0.08$ for $N = 34$). No relationships were found between the structure of an organization and the ownership pattern of its parent organization. This lack of relationship is quite striking, particularly in view of the extent of the correlation found with other contextual variables. Since ownership and control seemed to have its impact through the degree of public accountability, and the other variables did not have an additional effect, there seemed to be grounds for not proceeding with them in a multivariate analysis.

SIZE

There has been much work relating size to group and individual variables, such as morale and job satisfaction, with not very consistent results (Porter and Lawler, 1965). With few exceptions, empirical studies relating size to variables of organization structure have confined themselves to those broad aspects of the role structure which are here termed "configuration" (Starbuck, 1965). Hall and Tittle (1966), using a Guttman scale of the overall degree of perceived bureaucratization obtained by combining scores on six dimensions of Weberian characteristics of bureaucracy in a study of twenty-five different work organizations, found a small relation between their measurement of perceived bureaucratization and organization size ($\tau = 0.252$ at the 6 percent level of confidence).

In this study the aspects of size studied were number of employees, net assets utilized, and number of employees in the parent organizations.

Number of Employees and Net Assets. It was intended that the sample be taken from the population of work organizations in the region employing more than 250 people, but the sample ranges from an insurance company employing 241 people to a vehicle manufacturing company employing 25,052 (mean 3,370; standard deviation 5,313). In view of this distribution, it was felt that a better estimate of the correlation between size and other variables would be obtained by taking the logarithm of the number of employees (mean 3.12; standard deviation 0.57).

"Net assets employed by the organization" was also used, because financial size might expose some interesting relationships with organization structure that would not appear when only personnel size was considered. The sample ranged from under £100,000—an estimate for the government inspection agency whose equipment was provided by its clients—to a confectionery manufacturing firm with £38 million. The attempt to differentiate between these two aspects of size proved unsuccessful, however, as the high correlation between them ($r = 0.78$) shows. Taking the logarithm of the two variables raised the correlation ($r = 0.81$). For this sample, therefore, a large organization was big both in number of employees and in financial assets. The logarithm of the number of employees was therefore taken to represent both these aspects of size.

The correlation between the logarithm of size and structuring of activities ($r = 0.69$) lends strong support to descriptive studies of the effects of size on bureaucratization. (This correlation may be compared with that between actual size and structuring of activities, $r = 0.56$, to demonstrate the effects of the logarithmic transformation.) Larger organizations tend to have more specialization, more standardization and more formalization than smaller organizations. The *lack* of relationship between size and the remaining structural dimensions, i.e., concentration of authority ($r = -0.10$) and line control of workflow ($r = -0.15$) was equally striking. This clear differential relationship of organization size to the various structural dimensions underlines the necessity of a multivariate approach to context and structure if oversimplifications are to be avoided.

Indeed, closer examination of the relationship of size to the main structural variables underlying the dimension of concentration of authority (Pugh, *et al.*, 1968: Table 4) points up a limitation in the present approach, which seeks to establish basic dimensions by means of factor analysis. As was explained in that paper, the structural factors represent an attempt to summarize a large amount of data on a large number of variables to make possible empirically based comparisons. But the cost is that the factor may obscure particular relationships with the source variables which it summarizes. For some purposes therefore, it may be interesting to examine particular relationships. The lack of relationship between size and concentration of authority, for example, summarizes (and therefore conceals) two small but distinct relationships with two of the component variables. There is no relationship between size and autonomy ($r = 0.09$), but there is a negative relationship between size and centralization ($r = -0.39$), and a positive one between size and standardization of procedures for selection and advancement ($r = 0.31$). The relationship with centralization has clear implications for the concept of bureaucracy. Centralization correlates *negatively* with all scales of structuring of activities except one: the more specialized, standardized, and formalized the organization, the *less* it is centralized. Therefore on the basis of these scales, there can be no unitary bureaucracy, for an organization that develops specialist offices and associated routines is decentralized. Perhaps when the responsibilities of specialized roles are narrowly defined, and activities are regulated by standardized procedures and are formalized in records, then authority can safely be decentralized. Pugh, *et al.* (1969) discuss the interrelationship of the structural variables in particular types of organization.

Size of Parent Organization. This is the number of employees of any larger organization to which the unit belongs. The literature on bureaucracy often implies that it is the size of the larger parent organization that influences the structure of the sub-unit. The important factor about a small government agency may not be its own size, but that of the large ministry of state of which it is a part. Similarly, the structure of a subsidiary company may be more related to the size of its holding company. The number of employees in the parent organizations ranged from 460 to 358,000 employees. The size of the parent organization correlated positively (after logarithmic transformation) with structuring ($r = 0.39$) and concentration of authority ($r = 0.39$) but not with line control of workflow ($r = -0.07$). The classical concept of bureaucracy would lead to the hypothesis that the size of the parent organization would be highly correlated with structuring of activities and concentration of authority, therefore the support from this sample was relatively modest. The correlation with structuring ($r = 0.39$) is much lower than the correlation of *organization* size and structuring ($r = 0.69$). The impact of the size of an organization is thus considerably greater than the size of the parent organization on specialization, standardization, formalization, etc. But a relationship with concentration of authority is not found with organization size ($r = -0.10$). Thus large groups have a small but definite tendency to have more centralized subunits with less autonomy. This relationship would be partly due to the government-owned organizations, inevitably part of large groups, which were at the concentrated end of this factor.

CHARTER

Scales. Institutional analysts have demonstrated the importance of the charter of an organization; that is, its social function, goals, ideology, and value systems, in influencing structure and functioning (Parsons, 1965; Selznick, 1949). To transform concepts which had been treated only descriptively into a quantitative form that would make them comparable to other contextual aspects, seven ordered category scales were devised. Four of them characterized the purpose or goal of the organization in terms of its "output," the term being taken as equally applicable to products or services: (1) multiplicity of outputs—ranging from a single standard output to two or more outputs, (2) type of output—a manufacturing-service dichotomy, (3) consumer or producer outputs or a mixture of both, and (4) customer orientation of outputs—ranging from completely standard outputs to outputs designed entirely to customer or client specification. Three scales were devised for ideological aspects of charter: (5) self-image—whether the ideology of the organization as indicated by slogans used and image sought emphasized the qualities of its outputs; (6) policy on multiple outputs—whether the policy was to expand, maintain, or contract its range of outputs; and (7) client selection—whether any, some, or no selectivity was shown in the range of customers or clients served by the organization. Table 4–6 gives the details of the seven scales and Table 4–7 the intercorrelation matrix between them. This suggested that factor analysis would be helpful in summarizing the data, and a principal components analysis applied to the matrix gave the results shown in Table 4–8.

Operating Variability. This factor, accounting for 30 percent of the variance was highly loaded on the variables, consumer or producer outputs, customer orientation of outputs, and type of output. It was therefore conceptualized as being concerned with manufacturing nonstandard producer goods as against providing standard consumer service. The manufacturing producer end of the scale was linked with an organizational emphasis on self-image, whereas the consumer service end emphasized outputs. The scale was therefore constructed by a weighted summing

TABLE 4-6
Charter

Distribution N = 46	Score	Scale number and title
		Scale No. 14.02
		Multiplicity of outputs
19	1	Single output with standard variations
8	2	Single output with variations to customer specification
19	3	Two or more outputs
		Scale No. 14.03
		Type of output
14	1	Service (nonmanufacturing)
32	2	Manufacturing (new physical outputs in solid, liquid or gaseous form)
		Scale No. 14.04
		Type of output
16	1	Consumer (outputs disposed of to the general public or individuals)
7	2	Consumer and producer
23	3	Producer (outputs disposed of to other organizations which use them for, or as part of, other outputs)
		Scale No. 14.06
		Customer orientation
11	1	Standard output(s)
7	2	Standard output(s) with standard modifications
6	3	Standard output(s) with modification to customer specification
22	4	Output to customer specification
		Scale No. 14.07
		Self-image
24	1	Image emphasizes qualities of the *organization itself*
6	2	Image emphasizes both the organization and the output
16	3	Image emphasizes qualities of the *output* of the organization
		Scale No. 14.08
		Policy on outputs multiplicity
5	1	Contracting the range of outputs
26	2	Maintaining the range
15	3	Expanding the range
		Scale No. 14.09
		Ideology: client selection
28	1	No selection, any clients supplied
14	2	Some selection of clients
4	3	Clients specified by parent organization

of the scores on all these variables (the weighting being necessary to equate the standard deviations) and then standardizing the sums to a mean of 50 and a standard deviation of 15. This produced the range of scores on the scale given in Table 4-9. The lower scores distinguished

TABLE 4-7
Intercorrelation Matrix (Product-moment Coefficients, $N = 46$)

	Multiplicity of outputs	Service-manufacturing	Consumer-producer	Customer orientation	Self-image	Client selection	Expansion-contraction of range
Multiplicity of outputs	–						
Service-manufacturing	0.15	–					
Consumer-producer	0.05	0.37	–				
Customer orientation	0.38	0.18	0.59	–			
Self-image	–0.05	–0.17	–0.33	–0.13	–		
Client selection	–0.14	–0.02	0.28	–0.04	–0.18	–	
Expansion-contraction of range	0.07	–0.14	–0.09	0.07	0.10	–0.09	–

organizations giving only a standard service (e.g., teaching, transport, retailing), from organizations (with high scores) producing nonstandard producer outputs to customer specification (metal goods firm, engineering repair unit, packaging manufacturer, etc.), with those organizations having a standard output range in the middle.

Operating Diversity. This factor of charter, accounting for 20 percent of the variance, emphasized multiplicity of outputs, policy on whether to expand the range of kinds of outputs, client selection, and self-image. The more diversely operating organizations were a glass manufacturer, a metal manufacturer, and a brewery; the more restricted were a motor component manufacturer, a domestic appliance manufacturer, and a scientific inspection agency.

Eisenstadt (1959), Parsons (1956), Selznick (1949, 1957), Wilson (1962), and Clark (1956) have discussed the effects of the goals of an organization on its structure, but there has been almost no detailed empirical work on the actual relationship between goals and structure. Selznick (1949) showed how the goal of democracy led to decentralization in the TVA, and also suggested that the role structure of an organization is the institutional embodiment of its purpose. Wilson (1962) suggested a relationship between goals and methods of recruitment and means of selection. Clark (1956) as well as Thompson and Bates (1957) emphasized both the marginality and the degree of concreteness of the goal as a determinant of the direction of organizational adaptation. Blau

TABLE 4-8
Charter: Principal-components Analysis

Scales	Factor loadings: Operating variability*	Factor loadings: Operating diversity†
Consumer-producer output	0.85	0.16
Customer orientation of outputs	0.74	−0.41
Type of output (service-manufacturing)	0.57	0.00
Self-image	−0.52	−0.34
Multiplicity of outputs	0.37	−0.66
Client selection	0.23	0.66
Expansion-contraction of range	−0.15	−0.48

*Percentage of variance = 30%.
†Percentage of variance = 20%.

and Scott (1962) made one of the few attempts to classify organizations by their goals, suggesting that internal democracy goes with mutual benefit goals, efficiency with business goals, a professional structure with service goals and bureaucratic structure with commonweal goals.

Scales of organizational charter were related to structure, and operating variability was shown to be strongly associated with line control of workflow ($r = -0.57$). Thus the more an organization is concerned with manufacturing nonstandard producer goods, the more it relies upon impersonal control of workflow; the more it is providing a standard consumer service, the more it uses line control of its workflow through the supervisory hierarchy. Organizations showing operating diversity, however, tended to be more structured in activities ($r = 0.26$) and more dispersed in authority ($r = -0.30$).

TECHNOLOGY

Scales. Technology has come to be considered increasingly important as a determinant of organizational structure and functioning, although comparative empirical studies of its effects on structure are few, mainly case studies on the effects on the operator's job and attitudes (Walker, 1962). Thompson and Bates (1957), however, compared a hospital, a university, a manufacturing organization, and a mine for the effects of their technologies on the setting of objectives, the management of resources, and the execution of policy. The main work on the classification of technology in relation to organization structure has been that of Woodward (1965). She related mainly "configuration" aspects of the structure of manufacturing organization (e.g., number of levels of authority, width of

TABLE 4-9
Operating Variability

Number of organizations N = 46	Score	Type of organization
1	48	Component manufacturer
6	45	Two metal goods manufacturers Component manufacturer Abrasives manufacturer Packaging manufacturer Glass manufacturer
2	43	Printer Repairs for government department
4	42	Two component manufacturers Motor component manufacturer Metal motor component manufacturer
2	41	Vehicle manufacturer Engineering tool manufacturer
1	40	Component manufacturer
3	37	Civil engineering firm Carriage manufacturer Metal goods manufacturer
3	36	Vehicle manufacturer Confectionery manufacturer Local authority water department
2	35	Motor-tire manufacturer Commercial vehicle manufacturer
4	34	Motor component manufacturer Non-ferrous metal manufacturer Research division Food manufacturer
3	33	Engineering component manufacturer Domestic appliances manufacturer Local authority civil engineering department
1	32	Component manufacturer
2	31	Government inspection department Toy manufacturer
3	30	Brewery Insurance company Food manufacturer
1	27	Local authority transport department
6	25	Local authority baths department Co-operative chain of retail stores Chain of retail stores Savings bank Chain of shoe repair stores Department store
1	23	Omnibus company
1	21	Local authority education department

spans of control) to a classification of their production systems according to the "controlability and predictability" of the process.

In the present study the need to develop suitable measurements of overall organizational technology made the level of generality achieved by the Woodward classification desirable; but the need to develop concepts of technology that applied to all the organizations in the sample precluded the direct adoption of that scale. A full account of the development of scales of technology and their relationship to organization structure is given in Hickson, *et al.* (1969). Only the scales included in the present analysis are described here.

Technology is here defined as the sequence of physical techniques used upon the workflow of the organization, even if the physical techniques involve only pen, ink, and paper. The concept covers both the pattern of operations and the equipment used, and all the scales developed are applicable to service as well as to manufacturing organizations. Five scales of related aspects of technology were developed.

Thompson and Bates (1957) defined the "adaptability" of the technology as "the extent to which the appropriate mechanics, knowledge, skills and raw materials can be used for other products" and, it may be added, services. An attempt to operationalize some aspects of this definition is given in Table 4–10, which shows a scale of *workflow rigidity*. This consists of eight biserial items concerned with the adaptability in the patterns of operations; for example, whether the equipment was predominantly multi-purpose or single-purpose, whether rerouting of work was possible, etc. Since this was a scale of composite items, item analysis was used to test the scaleability. The mean item analysis value of 0.84 indicates that it is legitimate to add the scores on these items to form a workflow rigidity score for an organization.

Two other scales of technology utilized the concepts outlined by Amber and Amber (1962). They postulated that "the more human attributes performed by a machine, the higher its automaticity" and compiled a scale of automaticity together with clear operational definitions, which could be applied to any piece of equipment from a pencil to a computer, and which categorized each into one of six classes. The two scales based on these concepts were: the *automaticity mode,* i.e., the level of automaticity of the bulk of the equipment of the organization; and the *automaticity range*; i.e., the highest-scoring piece of equipment an organization used, since every organization also scored the lowest possible by using hand tools and manual machines.

TABLE 4-10
Scale of Workflow Rigidity

Item	*Number of organizations (N = 52)*⁺	*Item analysis value**
No waiting time possible (versus waiting time)	8	0.82
Single-purpose equipment (versus multi-purpose)	13	0.78
Production of service line (versus no set line)	42	1.00
No buffer stocks and no delays possible (versus buffer stocks and delays)	8	0.71
Single-source input (versus multi-source input)	12	0.67
No rerouting of work possible (versus rerouting possible)	15	0.80
Breakdown stops all workflow immediately (versus not all workflow stops)	6	0.97
Breakdown stops some or all workflow immediately (versus no workflow stops)	35	0.95

*Mean item analysis value = 0.84.
⁺Since this is a test of internal consistency and scaleability, the whole group of 52 organizations was used. (D.S. Pugh *et al.*, 1968.)

The fourth scale, *interdependence of workflow segments*, was a scale of the degree of linkage between the segments of an organization; a segment being defined as those parts into which the workflow hierarchy was divided at the first point of division beneath the chief executive. The three points on the scale were: (1) segments duplicated in different locations, all having the same final outputs; (2) segments having different final outputs, which are not inputs of other segments; (3) segments having outputs which become inputs of other segments. The final scale, *specificity of criteria of quality evaluation*, was a first attempt to classify the precision with which the output was compared to an acceptable standard. The three points on the scale were: (1) personal evaluation only; (2) partial measurements, of some aspect(s) of the output(s); (3) measurements used over virtually the whole output, to compare against precise specification (the "blueprint" concept).

Correlations. As expected, these measures tend to be highly intercorrelated. A principal components analysis extracted a large first factor accounting for 58 percent of the variance, with loadings of over 0.6 on all scales, and of over 0.8 on three of them. A scale of *workflow integration* was therefore constructed by summing the scores on the component scales. Among organizations scoring high, with very integrated, automated, and rather rigid technologies, were an automobile factory, a food manufacturer, and a swimming baths department. Among those scoring low, with diverse, nonautomated, flexible technologies, were retail stores, an education department, and a building firm.

There were no clear relationships between workflow integration and the variables of size, origin and history, or concentration of ownership with control and negative relationship with public accountability ($r = -0.35$), largely because the government-owned organizations in the sample were predominantly service and therefore at the diverse end of the workflow integration scale. The correlations between workflow integration and operating variability ($r = 0.57$) and diversity ($r = 0.33$) reflect the close relationship between the ends of the organization and the means it employs to attain them.

Workflow integration showed modest but distinct correlations with all the three structural factors, the only contextual variable to do so, as can be seen from Table 4–2. The relationships of technology are therefore much more general than is the case with size, for example, which has a greater but more specific effect. The positive correlation between workflow integration and structuring of activities ($r = 0.34$) would be expected since highly integrated and therefore more rigid technologies would be associated with a greater structuring of activities and procedures. Similarly, the correlation with concentration of authority ($r = -0.30$) suggests that because of the increasing control resulting directly from the workflow itself in an integrated technology, decisions tend to become more routine and can be decentralized. But the fact that the correlations are not higher than this emphasizes that structuring may be related to other contextual factors, such as size. The relationship of technology to line control of workflow, however, was very clear ($r = 0.46$); the more integrated the technology, the more the reliance on impersonal control. It must be emphasized, however, that these relationships were found on the whole sample of manufacturing and service organizations. When manufacturing organizations only were considered, some of the relationships showed considerable change (Hickson *et al.*, 1969).

Labor Costs. This is a second related, but conceptually distinct, aspect of the technology of the workflow and is expressed as a percentage of total costs. The range in the sample was from 5 to 70 percent, with engineering organizations scoring low and public services high. The scale correlated with workflow integration ($r = -0.50$), high integration being associated with reduced labor costs. Its correlations with the structural factors are comparable with those for technology (after adjusting the signs).

LOCATION

The geographical, cultural, and community setting can influence the organization markedly (Blau and Scott, 1962). This study controls for some of these effects in a gross way, for all organizations of the sample were located in the same large industrial conurbation, and the community and its influence on the organizations located there were taken as given (Duncan *et al.*, 1963). Compared with the national distribution, the sample was overrepresented in the engineering and metal industries, and unrepresented in mining, shipbuilding, oil refining and other industries. Because of the location, however, regional cultural differences of the sort found by Thomas (1959) as to role conceptions, were avoided.

One aspect of location which discriminated between organizations in the sample, was *number of operating sites*. The range formed a Poisson distribution, with 47 percent of the sample having one site; but six organizations had over a hundred sites, and two over a thousand. This distribution did not appear to be a function of size ($r = 0.14$) but of the operating variability aspect of charter ($r = -0.56$). Manufacturing organizations were concentrated in a small number of sites (the largest number being nine), whereas services range across the scale. The number of operating sites was therefore correlated with the workflow integration scale of technology ($r = -0.58$), and with public accountability ($r = 0.34$), this last correlation reflecting the predominantly service function of the group of government-owned organizations.

This pattern of inter-relationships among the contextual variables led to the expectation of relationships between number of operating sites and the structural dimensions which would be congruent with those of operating variability and workflow integration. The correlations of number of operating sites with structuring of activities ($r = -0.26$), concentration of authority ($r = 0.39$) and line control of workflow ($r = 0.39$) confirm the relationships with charter and technology, and suggest a *charter-technology-location* nexus of interrelated contextual variables having a combined effect on structure.

DEPENDENCE

The dependence of an organization reflects its relationships with other organizations in its social environment, such as suppliers, customers, competitors, labor unions, management organizations, and political and social organizations.

Dependence on Parent Organization. The most important relationship would be the dependence of the organization on its parent organization. The *relative size* of the organization in relation to the parent organization was calculated as a percentage of the number of employees. This ranged from under one percent in two cases—a branch factory of the central government, and a small subsidiary company of one of the largest British private corporations in the country—to 100 percent in eight independent organizations. The distribution was Poisson in form with a mean and standard deviation of 37 percent. The next scale was a four-point category scale concerned with the *status* of the organization in relation to the parent organization: (1) principal units (8 organizations) where the organization was independent of any larger group although it might itself have had subsidiaries or branches; (2) subsidiary units (18 organizations) which, although part of a larger group, had their own legal identity with, for example, their own boards of directors; (3) head branch units (4 organizations) which did not have separate legal identity although they were the major operating components of the parent organization and the head office of the parent organization was on the same site; (4) branch units (16 organizations) operating parts of a parent organization which did not satisfy the preceding criteria.

The third aspect of the relation between the organization and the parent organization was given by the degree of *organizational representation on policy-making bodies.* This three-point scale ranged from the organization being represented on the policy-making body of the parent organization (e.g. board of directors, city council), through the organization being represented on an intermediate policy-making body (e.g. board of directors of an operating company but not of the ultimate owning holding company, committee of the city council), to the organization having no representative on any policy-making body of the parent organization. As would be expected, these three variables were highly correlated (Tables 4–11, 4–12).

A related variable was the number of *specializations contracted out* by the organization. In many cases these would be available as services of the parent organization to the organization, although account was also taken of the various specialist services (e.g., consultants) used outside the parent organization. The specializations were as defined in the structural scale of functional specialization (Pugh *et al.*, 1968: Appendix A), and ranged from one specialization contracted out (two engineering works, a printer, and a builder) to no less than fifteen of the sixteen

TABLE 4-11
Dependence

Distribution N = 46	Score	Scale number and title
		Scale No. 18.07
		*Relative size**
		Range = 0-100 Mean = 37.4 S.D. = 37.3
		Scale No. 12.10
		Status of organization unit
16	1	Branch
4	2	Head branch (headquarters on same location)
18	3	Subsidiary (legal identity)
8	4	Principal unit
		Scale No. 12.11
		Organizational representation on policy-making bodies†
19	1	Organization not represented on top policy-making body
4	2	Organization represented on local policy-making body but not on top policy-making body
23	3	Organization represented on policy-making body
		Scale No. 18.06
		Number of specializations contracted out‡
		Range = 1-16 Mean = 7.2 S.D. = 4.0
		Scale No. 18.17
		Vertical integration §
		Range = 1-16 Mean = 7.7 S.D. = 3.5
		Scale No. 18.03
		Integration with suppliers
4	1	No ownership ties and single orders
7	2	No ownership and single contracts or tenders
8	3	No ownership and short-term contracts, schedule and call-off
6	4	No ownership and yearly contracts, standing orders
7	5	Ownership and contractual ties
14	6	Ownership and tied supply
		Scale No. 18.05
		Response in outputs volume to customer influence
12	1	Outputs for stock
5	2	Outputs for stock and to customer order
21	3	Outputs to customer order
2	4	Outputs to customer order and to schedule and call-off
6	5	Outputs to schedule and call-off
		Scale No. 18.08
		Integration with customers: type of link with customers
24	1	Single orders
9	2	Regular contracts
10	3	Long-term contracts (over two years)
3	4	Ownership
		Scale No. 18.09
		Integration with customers: dependence of organization on its largest customer
30	1	Minor outlet (less than 10% of output)
10	2	Medium outlet (over 10% of output)
3	3	Major outlet (over 50% of output)
3	4	Sole outlet
		Scale No. 18.10
		Integration with customers: dependence of largest customer on organization
11	1	Minor supplier (less than 10% of particular item)
5	2	Medium supplier (over 10% of particular item)
21	3	Major supplier (over 50% of particular item)
9	4	Sole supplier with exclusive franchise

TABLE 4-12
Dependence: Intercorrelation Matrix
(Product-moment Coefficient, $N = 46$)

	Relative size	*Status of organization unit*	*Organizational representation on policy-making body*	*Specializations contracted out*	*Vertical integration*	*Trade unions*
Relative size	–					
Status of organization unit	0.68	–				
Organizational representation on policy-making body	0.50	0.65	–			
Specializations contracted out	−0.60	−0.51	−0.52	–		
Vertical integration	−0.40	−0.34	−0.36	0.45	–	
Trade unions	−0.09	−0.16	−0.25	0.19	0.28	–

specializations contracted out (an abrasives manufacturer and a packaging manufacturer) with a mean 7.2 and standard deviation 4.0.

Dependence on Other Organizations. The suppliers and customers or clients of the organization must also be considered. The operating function of the organization can be regarded as being the processing of inputs and outputs between supplier and client, and the degree to which the organization is integrated into the processual chain by links at either end can be measured. Five category scales were developed to elucidate this concept (with details given in Table 4–11). They were concerned with the integration with suppliers and clients, and response in the output volume to client influence, etc. To establish a single dimension measuring the degree to which the organization was integrated into this system, the

TABLE 4-11 FOOTNOTES

*Size of unit as a percentage of size of parent organization.

†Internal and parent organizations.

‡The specializations are those of functional specialization (D.S. Pugh *et al.*, 1968: Appendix A). Scores are out of a possible 16.

§This scale is formed by the total of the scores on the 18 items representing the following five scales: 18.03, 18.05, 18.08, 18.09, 18.10.

five scales were transformed into biserial form. Item analysis was carried out on the 18-item scale generated and yielded a mean item analysis value of 0.70, which seemed to justify the addition of the items into a total scale, *vertical integration*. At one extreme was a confectionery manufacturer and an engineering components firm supplying goods from stock with a large number of customers after obtaining their supplies from a large variety of sources; at the other extreme were organizations (vehicle components, civil engineering, scientific research) obtaining their resources from a small number of suppliers and supplying their product or service to a small number of clients (often the owning group only) who had a marked effect upon their workflow scheduling.

For *trade unions*, a scale of five ordered categories was developed of the extent to which unions were accepted as relevant to the activities of the organization. The scale was (1) no recognition given; (2) only partial recognition given (i.e., discussions for certain purposes, but not negotiations); (3) full recognition given to negotiate on wages and conditions of service on behalf of their members; (4) full recognition given plus facilities for union meetings to be held regularly on the time and premises of the organization; (5) as in the preceding plus the recognition of a works convenor to act on behalf of all unions with members in the organization. Organizations in the sample were located in all the categories, with the modal position being full recognition; but five organizations did not recognize unions, and eleven gave the maximum recognition including a works convenor.

Examination of the intercorrelations between these six variables of dependence (Table 4–12) and of their correlations with other important aspects of context (Table 4–13) shows considerably higher correlations with size of parent organization than with size of organization, and considerably higher correlations with concentration of ownership with control (a variable applied to the parent organization) than with operating variability or workflow integration (variables applied to the operations of the individual organizations themselves). This pattern lends support to the view that these measures are tapping aspects of the dependence of the organization, particularly its dependence on external resources and power as in Eisenstadt's (1959) formulation. The one exception was the variable of recognition of trade unions, which had its largest contextual correlation with organization size, and is therefore concerned with a different aspect of interdependence. Impersonality of origin (from origin and history) and public accountability (from ownership and control) show the same pattern of higher correlation with the parent organization than with the

TABLE 4-13
Dependence

	Concentration of ownership with control	*Size*	*Size of parent organization*	*Operating variability*	*Workflow integration*	*Size of organization*	*Structuring of activities*	*Concentration of authority*	*Line control of workflow*
Status of organization unit	0.45	0.11	−0.27	0.01	0.05	0.17	0.13	−0.63	−0.07
Organizational representation on policy-making bodies	0.41	0.15	−0.19	−0.01	0.19	0.20	0.14	−0.63	−0.18
Number of specializations contracted out	−0.32	0.14	0.40	0.11	0.09	0.01	0.18	0.53	0.00
Relative size	0.47	0.06	−0.38	−0.08	−0.03	0.16	0.03	−0.40	−0.13
Vertical integration	−0.15	−0.01	0.39	0.21	−0.12	−0.06	0.06	0.29	−0.04
Trade unions*	−0.21	0.26	0.25	0.19	0.21	0.36	0.51	0.08	−0.35
Impersonality of origin	−0.40	0.13	0.36	−0.27	−0.25	0.07	−0.04	0.64	0.36
Public accountability of parent organization	−0.51	0.04	0.45	−0.35	−0.35	0.01	−0.10	0.64	0.47
Dependence	−0.49	−0.06	0.37	0.05	−0.05	−0.17	−0.05	0.66	0.13

*This variable was not included in the scale of dependence

unit, indicating that impersonally founded organizations are likely to be more dependent on their founding organizations; and that more publicly accountable organizations are more likely to be dependent on outside power with government-owned organizations being the extreme case.

These relationships suggested the application of factor analysis to a correlation matrix containing the seven variables. A principal-components analysis applied to the matrix produced a large first factor *dependence*[1] accounting for 55 percent of the variance, which was heavily loaded on all seven scales (on six of the seven, the loadings were above 0.7; the remaining loading on vertical integration was 0.58). The scores for dependence were obtained by an algebraic weighted sum of the scores on the four most highly loaded component scales, the weightings being obtained by a multiple regression analysis of the component scales on the factor. A high score characterized organizations with a high degree of dependence, which tended to be impersonally founded, publicly accountable, vertically integrated, with a large number of specializations contracted out, small in size relative to their parent organization, low in status, and not represented at the policy-making level in the parent organization (e.g., branch units in packaging, civil engineering, and food manufacture, a central government repair department, and a local government baths department). Organizations with low dependence were independent organizations characterized by personal foundation, low public accountability, little vertical integration, few specializations contracted out, and where the parent organization was the organization itself (e.g., a printing firm, the very old metal goods firm, a chain of shoe repair stores, and an engineering component manufacturer).

The correlation of dependence with the structural factors was focused largely on concentration of authority ($r = 0.66$), in every case, for dependence and its component scales the correlation being much greater than with the other factors, as Table 4–13 shows. Indeed, apart from the correlations with impersonality of origin and public accountability, none of the other correlations reached the 5 percent level of confidence. Dependent organizations have a more centralized authority structure and less autonomy in decision making; independent organizations have more autonomy and decentralize decisions down the hierarchy.

The relationships between dependence and the component scales of concentration of authority vary. Centralization, as defined and measured in this study, is concerned only with the level in the organization which has the necessary authority to take particular decisions, (Pugh *et al.*, 1968: 76); the higher the necessary level, the greater the centralization.

No account was taken of the degree of participation or consultation in decision-making as in Hage and Aiken's (1967) formulation of the concept. These were regarded as aspects for study at the group level of analysis. Neither is it possible for such a statement as the following to hold: "The decisions were centralized on the foreman since neither the superintendent nor the departmental manager had the necessary experience." In the present formulation this would be regarded as relative decentralization. Autonomy was measured by the proportion of decisions that could be taken within the organization as distinct from those which had to be taken at the level above it. Thus independent organizations of necessity had more autonomy, since there was no level above the chief executive, and the correlation between dependence and this component of concentration of authority was $r = -0.72$. The relation of centralization (which is concerned with the whole range of levels in the hierarchy) with dependence is less, but still high ($r = 0.57$). Dependent organizations also have a distinct tendency to standardize the procedures for selection and advancement ($r = 0.40$), a major component of concentration of authority. So dependent units have the apparatus of recruitment routines, selection panels, formal establishment figures, etc. of their parent organizations.

RELATION BETWEEN STRUCTURE AND CONTEXT

In this investigation of the relationship of organization structure to aspects of the context in which the organization functions, the use of scaling and factor analytic techniques has made possible the condensation of data and reorganization of concepts and has established eight distinctive scales of elements of context. These scales, shown in Table 4–14 together with their correlations with the structural dimensions, denote the variables that are salient among those which have been thought to affect structure. Relationships between structure and age, size, charter (operating variability, operating diversity), technology (workflow integration) location (number of operating sites) and dependence on other organizations are exposed by the correlations. At the same time the correlations raise questions about the relationship between ownership pattern and administrative structure.

The Multivariate Prediction of Structure from Context

From inspection of Table 4–14 and of the intercorrelation matrix in Table 4–3, certain elements of context can now be identified. The variables in Table 4–14 are now used as independent variables in a prediction analysis of the structural dimensions. The pattern of these correlations, that is, that where they are high they are specific, and where they are low, they are diffused, indicates that the predictions should be attempted on a multivariate basis. In this case consideration had to be given to choosing not only predictors with high correlations with the criterion, but also having low intercorrelations among themselves. If high intercorrelations among the predictors were allowed, then, since the high correlations with the criterion would be aspects of the same relationship, the multiple correlation would not be increased to any extent. If the intercorrelations between the predictors were low, then each would make its distinct contribution to the multiple correlation.

These problems can be illustrated from the attempt to obtain a multiple prediction of structuring of activities from the three contextual variables correlated with it (Table 4–14). Size is clearly the first predictor, with a correlation of $r = 0.69$, and the question is whether taking account of size of parent organization and workflow integration will increase pre-

TABLE 4-14
Salient Elements of Context (Product-moment Correlations with Structural Factors)*

Elements of context	*Structuring of activities*	*Concentration of authority*	*Line control of workflow*
Age	–	–0.38	–
Size of organization†	0.69	–	–
Size of parent organization†	0.39	0.39	–
Operating variability	–	–	–0.57
Operating diversity	–	–0.30	–
Workflow integration	0.34	–0.30	–0.46
Number of operating sites	–	0.39	0.39
Dependence	–	0.66	–

*With $N = 46$, correlations of 0.29 are at the 5% level of confidence, and correlations of 0.38 are at the 1% level of confidence.

†Logarithm of number of employees.

dictive accuracy. In spite of its greater correlation with the criterion, the size of the parent organization would be expected to make a smaller contribution to the prediction that workflow integration, since it has a strong correlation with the first predictor ($r = 0.43$); whereas the technology measure is not correlated with organization size ($r = 0.08$). This is in fact the case as shown in the first section of Table 4–15, which gives the multiple prediction analyses for the three structural factors.

Table 4–15 shows for each predictor variable, the single correlation with the criterion, the multiple correlation obtained by adding this predictor to the preceding ones, the F ratio corresponding to the increase obtained on the addition of this predictor, the degrees of freedom corresponding to the F ratio when $N = 46$, and the level of confidence at which the increase due to this predictor can be quoted. It will be seen from the first section of Table 4–15 that the correlation 0.69, between size and structuring of activities, is increased to a multiple correlation of 0.75 when workflow integration is added as a predictor. But the multiple correlation shows no noticeable increase when size of parent organization is added as a third predictor; that is, its predictive power has already been tapped by the two previous variables.

It must be emphasized that this procedure assesses only the predictive power of the contextual variables, not their relative importance in any

TABLE 4-15
Multiple Prediction Analysis of Structural Factors

Contextual predictors of structural factors	*Single correlation*	*Multiple correlation*	*F ratio*	*Degrees of freedom*	*Level of confidence*
Structuring of activities					
Size	0.69	0.69	39.6	1 : 44	>99%
Workflow integration	0.34	0.75	8.2	1 : 43	>99%
Size of parent organization	0.39	0.76	1.9	1 : 42	NS
Concentration of authority					
Dependence	0.66	0.66	34.2	1 : 44	>99%
Location (number of operating sites)	0.39	0.75	12.5	1 : 43	>99%
Age of organization	−0.38	0.77	2.5	1 : 42	NS
Operating diversity	−0.30	0.78	3.0	1 : 41	NS
Workflow integration	−0.30	0.78	0.0	1 : 40	NS
Size of parent organization	0.39	0.79	0.4	1 : 39	NS
Line control of workflow					
Operating variability	−0.57	0.57	20.7	1 : 44	>99%
Workflow integration	−0.46	0.59	1.7	1 : 43	NS
Number of sites	0.39	0.59	0.1	1 : 42	NS

more general sense. It cannot be concluded that the relationship of size of parent organization to structuring of activities is less important than that of workflow integration, because it adds less to the multiple correlation. Indeed the original higher correlation shows that this is not the case. Because of the interaction of the variables, the effects of organizational size and size of parent organization are confounded, in this study, as the correlation between them shows. A full examination of their relative effects would require a sample in which they were not correlated, as is the case with the technology measure.

The same argument applies to the multiple prediction of concentration of authority (Table 4–15). Here again there is a clear first predictor, dependence, with a correlation of 0.66 but then a choice of intercorrelated variables. The selection was made in order to get as high a multiple correlation as possible with as few predictors as possible, but the fact that the later predictors add nothing to the multiple correlation does not mean that they have no impact, only that predictive power has been exhausted by previous related variables. The existence of the charter-technology-location nexus referred to above is supported by the fact that when any one of these variables is used as a predictor, the remaining two do not add to the multiple correlation. Table 4–15 shows the multiple correlation of 0.75 obtained by using the location measure together with dependence as predictors. When the technology scale of workflow integration is substituted as the second predictor, the multiple correlation is 0.71; when the operating diversity scale of charter is used, the multiple correlation is 0.70.

The prediction of line control of workflow shows this same phenomenon, where the addition of predictors, because of their inter-relationships, does not improve on the original single correlation of operating variability with the criterion.

The size of the multiple correlations obtained with the first two factors, each 0.75 with two predictors, together with the small number of predictors needed, strongly supports the view that in relation to organization structure as defined and measured in this study, salient elements of context have been identified. Thus a knowledge of the score of an organization on a small number of contextual variables makes it possible to predict within relatively close limits, its structural profile. Given information about how many employees an organization has, and an outline of its technology in terms of how integrated and automated the work process is, its structuring of activities can be estimated within fairly close limits. Since in turn the score of the organization on structuring of activities summarizes

TABLE 4-16
Multiple Regression on Structural Factors

Structural factors	Whole sample	Subsamples 1	Subsamples 2	"Robust" weightings
Structuring of activities				
Weightings of predictors				
Size	0.67	0.72	0.60	2
Workflow integration	0.29	0.14	0.43	1
Multiple correlation	0.75	0.73	0.79	0.74
Concentration of authority				
Weightings of predictors				
Dependence	0.64	0.50	0.77	2
Location	0.36	0.40	0.33	1
Multiple correlation	0.75	0.66	0.84	0.75

an extensive description of broad aspects of bureaucratization, the organization is thereby concisely portrayed in terms of this and similar concepts. Likewise, knowing the dependence of an organization on other organizations and its geographical dispersion over sites tells a great deal about the likely concentration of authority in its structure. *Size, technology, dependence and location* (number of sites) are critical in the prediction of the two major dimensions (structuring of activities, concentration of authority) of the structures of work organizations.

Multiple predictions of the order of magnitude obtained are as high as can be expected with this level of analysis. Higher values would imply that there were no important deviant cases, and that differences as to policies and procedures among the members of an organization have no effect on its structure. And this is obviously not so. The multiple predictions discussed here are applicable only to this sample. When the regression equations obtained are applied to another similar sample for prediction purposes, there is likely to be a reduction in the multiple correlations. The extent of this reduction can be strictly gauged only by investigating another similar sample of organizations. This cross-validation study is at present being undertaken, but a first attempt to estimate the likely amount of reduction was made by splitting the sample into two subsamples of 23 organizations, each stratified in the same way as the whole sample. Table 4–16 gives the multiple regressions on structuring of activities and concentration of authority for the whole sample and for the two subsamples separately. The multiple correlations and the weightings are of the same order of magnitude. A "robust" prediction on the basis of simple weightings was also calculated. These correlations should be less subject to shrinkage. The stability of the correlation of 0.57 between operating variability and line control of workflow is indicated by correlations on the two subsamples of 0.50 and 0.65.

Summary and Discussion

This study has demonstrated the possibilities of a multivariate approach to the analysis of the relationships between the structure of an organization and the context in which it functions. Starting from a framework as outlined in the conceptual scheme summarized in Table 4–1, aspects of the context and structure of the organization were sampled in order to establish scales which discriminated among organizations in a large number of aspects. From this sampling 103 primary scales of structure and context were developed as a basis for the analysis of the inter-relationships among them.

By scaling and factor analytic techniques, these were then summarized to form three basic dimensions of structure and eight salient elements of context (Table 4–14). The analogy with the psychological test constructor who samples behavior in order to establish dimensions of personality is clear, and the same limitations apply. Thus while a claim can be made for the internal consistency and scaleability of these measures, no claim can be made as to the comprehensiveness with which they cover the field. This is particularly clear in the attempt to elucidate aspects of context, a concept which, although in some respects narrower than that of environment, is still very wide. Emphasis was therefore placed on those aspects of context that had been held to be relevant to structure on the basis of previous writings. The size of the multiple correlations obtained indicates that at least some of the salient aspects of context were tapped.

The predicability of the structural dimensions from contextual elements serves as external validating evidence for the structural concepts themselves. It has now been shown that besides being internally consistent and scaleable, as previously demonstrated, they can also be related in a meaningful way to external referents. Indeed the size of the correlations inevitably raises the question of causal implications. It is tempting to argue that these clear relationships are causal—in particular, that *size, dependence, and the charter-technology-location nexus largely determine structure.*

It can be hypothesized that size causes structuring through its effect on intervening variables such as the frequency of decisions and social control. An increased scale of operation increases the frequency of recurrent events and the repetition of decisions, which are then standardized and formalized (Haas and Collen, 1963). Once the number of positions and people grows beyond control by personal interaction, the organization must be more explicitly structured. In so far as structuring includes the concept of bureaucracy, Weber's observation that "the increasing bureaucratic organization of all genuine mass parties offers the most striking example of the role of sheer quantity as a leverage for the bureaucratization of a social structure" is pertinent (Gerth and Mills, 1948).

Dependence causes concentration of authority at the apex of publicly owned organizations because pressure for public accountability requires the approval of central committees for many decisions. The similar position of small units in large privately owned groups is demonstrated by the effect that a merger may have upon authority. After a merger, a manager of the smaller unit "may no longer be able to take a certain decision and act upon it independently. He may have to refer matters to people who were complete strangers to him a few months earlier" (Stewart *et al.*, 1963).

Integrated technology may be hypothesized to cause an organization to move towards the impersonal control end of the line-control dimension. Line control is adequate in shops or in municipal schools or building maintenance gangs, where the technology of the tasks is not mechanized and each line supervisor and primary work group is independent of all the others. But as workflow integration reaches the production line or automated stages, where large numbers of tasks are interdependent, more control is needed than can be exercised by the line of command alone. Udy (1965) summarizes this in his proposition, "The more complex the technology . . . , the greater the emphasis on administration."

The causal argument need not run only one way. It can be suggested that a policy of specializing roles and standardizing procedures, that is, of structuring, would require more people, that is, growth in size. Concentration of decisions in the hands of an owning group is likely to result in more economic integration among the subsidiaries concerned, that is, more dependence; while the production control, inspection, and work-study procedures of staff control might raise the level of workflow integration in the technology.

But a cross-sectional study such as this can only establish relationships. Causes should be inferred from a theory that generates a dynamic model about changes over time. The contribution of the present study is to establish a framework of operationally defined and empirically validated concepts, which will enable processual and dynamic studies to be carried out on a much more rigorous and comparative basis than has been done previously. The framework is also seen as a means of controlling for organizational factors when individual and group level variables are being studied. Such studies must now be conducted with reference not only to differences in size, but also in dependence, operating function, workflow integration, etc., and with reference to the demonstrated relationship between these aspects of context and organization structure.

NOTE

1. We are grateful to our colleague Diana C. Pheysey for suggesting this formulation and for much valuable critical comment on an earlier draft of this paper.

REFERENCES

Aaronovitch, S.
1961 The Ruling Class. London: Lawrence and Wishart.
Amber, G. H., and P. S. Amber
1962 Anatomy of Automation. Englewood Cliffs, N.J.: Prentice-Hall.
Bakke, E. W.
1959 "Concepts of the social organization." In M. Haire (ed.). Modern Organization Theory. New York: Wiley.
Berle, A. A., and G. Means
1937 The Modern Corporation and Private Property. New York: Macmillan.
Blau, P., and W. R. Scott
1962 Formal Organizations. San Francisco: Chandler.
Brogden, H. E.
1949 "A new coefficient: application to biserial correlation and to estimation of selective efficiency." Psychometrika, 14: 169–182.
Burnham, J.
1962 The Managerial Revolution. London: Penguin.
Chandler, A. D.
1962 Strategy and Structure. Cambridge, Mass.: Massachusetts Institute of Technology.
Clark, B. R.
1956 "Organizational adaptation and precarious values." American Sociological Review, 21: 327–336.
Dahrendorf, R.
1959 Class and Conflict in Industrial Society. Stanford, Calif.: Stanford University
Dubin, R.
1958 The World of Work. Englewood Cliffs, N.J.: Prentice-Hall.
Duncan, O. D., W. R. Scott, S. Lieberson, B. Duncan and H. Winsborough
1963 Metropolis and Region. Baltimore, Md.: Johns Hopkins University.

Eisenstadt, S. N.
1959 "Bureaucracy, bureaucratization and debureaucratization." Administrative Science Quarterly, 4: 302–320.

Florence, P. S.
1961 Ownership, Control and Success of Large Companies. London: Sweet and Maxwell.

Gerth, H. H., and C. Wright Mills (eds.)
1948 From Max Weber: Essays in Sociology. London: Routledge and Kegan Paul.

Haas, E., and L. Collen
1963 "Administrative practices in university departments." Administrative Science Quarterly, 8: 44–60.

Hage, J., and M. Aiken
1967 "Relationship of centralization to other structural properties." Administrative Science Quarterly, 12 (June): 72–92.

Hall, R. H., and C. R. Tittle
1966 "A note on bureaucracy and its correlates." American Journal of Sociology, 72: 267–272.

Hickson, D. J., D. S. Pugh, and D. C. Pheysey
1969 "Operations technology and organizational structure: an empirical reappraisal." Administrative Science Quarterly, 14: 378–397.

Hinings, C. R., D. S. Pugh, D. J. Hickson, and C. Turner
1967 "An approach to the study of bureaucracy." Sociology, 1 (January): 62–72.

Inkson, J. H. K., R. L. Payne, and D. S. Pugh
1967 "Extending the occupational environment: the measurement of organizations." Occupational Psychology, 41: 33–47.

Levy, P., and D. S. Pugh
1969 "Scaling and multivariate analyses in the study of organizational variables." Sociology, 3 (May): 192–213.

Mills, C. Wright
1956 The Power Elite. London: Oxford University.

Parsons, T.
1956 "Suggestions for a sociological approach to the theory of organizations, I. and II." Administrative Science Quarterly, 1 (June and September): 63–85, 225–239.

Perrow, C.
1967 "A framework for the comparative analysis of organizations." American Sociological Review, 32 (April): 194–208.

Porter, L. W., and E. E. Lawler III
1965 "Properties of organization structure in relation to job attitudes and job behavior." Psychological Bulletin, 64: 25–51.

Presthus, R. V.
1958 "Towards a theory of organizational behavior." Administrative Science Quarterly, 3: 48–72.

Pugh, D. S., D. J. Hickson, and C. R. Hinings
1969 "An empirical taxonomy of work organization structures." Administrative Science Quarterly, 14 (March): 115–126.

Pugh, D. S., D. J. Hickson, C. R. Hinings, K. M. Macdonald, C. Turner, and T. Lupton
1963 "A conceptual scheme for organizational analysis." Administrative Science Quarterly, 8: 289–315.

Pugh, D. S., D. J. Hickson, C. R. Hinings, and C. Turner
1968 "Dimensions of organization structure." Administrative Science Quarterly, 13: 65–105.

Selznick, P.
1949 T.V.A. and the Grass Roots. Berkeley, Calif.: University of California.
1957 Leadership in Administration. Evanston, Ill.: Row and Peterson.
Starbuck, W. H.
1965 "Organizational growth and development." In J. G. March (ed.), Handbook of Organizations. Chicago: Rand McNally.
Stewart, R., P. Wingate, and R. Smith
1963 Mergers: The Impact on Managers. London: Acton Society Trust.
Stinchcombe, A. L.
1965 "Social structure and organization." In J. G. March (ed.), Handbook of Organizations. Chicago: Rand McNally.
Thomas, E. J.
1959 "Role conceptions and organizational size." American Sociological Review, 24: 30–37.
Thompson, J. D., and F. E. Bates
1957 "Technology, organization and administration." Administrative Science Quarterly, 2: 323–343.
Trist, E. L., G. W. Higgin, H. Murray, and A. B. Pollock
1963 Organizational Choice. London: Tavistock.
Udy, S. H.
1965 "The comparative analysis of organizations." In J. G. March (ed.), Handbook of Organizations. Chicago: Rand McNally.
Walker, C. R.
1962 Modern Technology and Civilization. New York: McGraw-Hill.
Wilson, B. R.
1962 "Analytical studies of social institutions." In A. T. Welford *et al.*, Society: Problems and Methods of Study. London: Routledge and Kegan Paul.
Woodward, J.
1965 Industrial Organization, Theory and Practice. London: Oxford University.

5

MARK LEFTON AND WILLIAM R. ROSENGREN

ORGANIZATIONS AND CLIENTS: LATERAL AND LONGITUDINAL DIMENSIONS*

This paper sets forth a framework for the development of an analytic model of formal organizations which views the clients of organizations as integral factors influencing the structure and functioning of such systems. Three reasons are basic to this effort. Of first importance is the fact that our age is witness to a new phase in the organizational revolution, one which is marked by a phenomenal growth in the number, variety, and importance of formal organizations which serve people as persons, rather than catering exclusively to material needs and wishes. Second, this transition appears to involve a major shift in the criteria by which the operations of organizations must be evaluated.[1] That is, the substitution of what may be called "humanitarian" values for purely economic and administrative considerations will eventually demand organizational responsiveness to an ethic of service rather than to one of efficiency. The third consideration has to do with the fact that emphasis upon such issues as rational efficiency, internal structures of authority and control, and the maintenance of organizational autonomy, while of obvious importance for the sake of better understanding economic and administrative organizations, may be of less utility in the analysis of organizations concerned with

Reprinted from Mark Lefton and William R. Rosengren, "Organizations and Clients: Lateral and Longitudinal Dimensions," *American Sociological Review*, 31–6 (December, 1966), pp. 802–881.

* We would like to thank Irving Rosow, Charles Perrow, and Robert Habenstein for offering us a number of helpful comments and criticisms. Many of their suggestions will also be incorporated in the further elaboration of the lateral-longitudinal model in a forthcoming monograph by William R. Rosengren and Mark Lefton, *Hospitals and Patients: A Theory of Clients and Organizations*.

the social and personal dilemmas of men. In addition, the existing conceptions of organizations may have to be broadened to cope with the interorganizational demands engendered by large-scale action programs in the fields of human welfare.

There are four distinct traditions in organizational analysis, none of which, for different reasons, has yet codified the linkages between clients and formal organizational structure. The first of these is perhaps best represented by the work which owes its principal intellectual debt to Weber's original conception of bureaucracy as a form of legitimate authority.[2] The second includes those studies dealing with the impact of the demographic and ecological characteristics of the surrounding community upon the formal structure and functioning of organizations.[3] Third, the social system perspective focuses more upon the structural linkages by which the functional requisites of formal organizations—seen as subsystems—are integrated with and accommodated to the institutional systems of the larger social order.[4] Finally, the fourth tradition in the study of organizations is represented by the symbolic interactionists.[5]

A diversity of organizational contexts have been examined from the perspective of Weber's ideal type, but the central focus has remained consistent with the bureaucratic model. The prime concern has been with the operating functionaries of organizations rather than with the clients they serve. The structural approach has yielded a large body of literature which compares and contrasts the formal properties of organizations with expectations derived from the ideal type.[6]

In contrast, the community structure approach has tended to focus either on cooptation, competition and other processes by which "publics in contact" are made congruent with organizational needs, or upon the manner in which organizations emerge as the demographic and ecological products of the host community.[7] A variant of this approach focuses specifically on the need for an organization to manipulate its incentive system in order to maintain the commitment of its members. Effort along these lines has attempted to link particular types of organizational incentive to varying "publics" and emphasize the strategic importance of examining an organization's sensitivity to changing motives as well as to environmental conditions.[8]

The social system approach, in its focus upon the systemic relations between organizations and the institutional sub-systems of which they are but a part, has tended to preclude a deliberate concern with the role of clients in organizations, precisely because of the level of analysis at which such concerns are generally expressed.

Finally, the symbolic interactionist approach leads to a conception of formal organizational structures and processes as having only secondary importance, providing only a contextual backdrop against which processes of self-identity, situation definitions, role emergence, and symbol verification are brought into bold relief.[9]

This brief discussion is designed to make one point: Insofar as they do not explicitly deal with the clients of organizations, the major traditions in organizational analysis remain conceptually divergent and substantially distinct.

Attempts to Relate Clients to Organizations

Congruent with the concerns indicated above, there has recently been an increased awareness on the part of several students of formal organizations of the need to regard clients as critical factors in organizational structure and functioning. For example, Parsons states: ". . . in the case of professional services there is another very important pattern where the recipient of the service becomes an operative member of the service-providing organization. . . . This taking of the customer *into* the organization has important implications for the nature of the organization."[10] But then the discussion is directed once again to a systemic analysis of the strategies by which organizations meet system-maintenance requisites without pursuit of the implications of the previous insight.

Blau and Scott explicitly identify some of the instances in which organizations might better be understood in the light of client characteristics. They say:

> It is perhaps a truism to say that organizations will reflect the characteristics of the publics they serve. A technical high school differs in predictable ways from a college preparatory school, and an upper-middle class church is unlike the mission church of the same denomination in the slums. While such differences seem to be important and pervasive, there has been little attempt to relate client characteristics systematically to organizational structures.[11]

It should be obvious that clients may present organizations with a wide range of characteristics. Any specific clientele characteristic may have a varying impact upon organizational functions, but only if such a characteristic is regarded as relevant. In this regard, organizations must select and define those client characteristics which are salient for their purposes. In a discussion of hospital structures, Perrow argues that hospitals belong to that class of organizations which attempt as their primary goal the alteration of the state of human material—such material being at once self-activating, subject to a multitude of orientations, "encrusted with cultural definitions," and embodying a wide range of organizational relevancies. Perrow then indicates the impacts that contrasting definitions of the client material are likely to have upon the technologies employed in hospitals and of their structural properties as well.[12]

The importance of contrasting definitions by organizations of the publics they serve, particularly for internal as well as external control processes, has been emphasized by Etzioni. From his perspective, a critical dimension in this respect derives from the confrontation between service as an ideology and service as an organizational instrument of manipulation and control.[13]

Another hint at the importance of clients is provided by Eisenstadt's discussion of debureaucratization. The client is here perceived as a scarce resource—a fact having implications for internal structure as well as for inter-organizational relationships. To the extent that the client does constitute a scarce resource upon which organizational survival depends, "the more (the organization) will have to develop techniques of communication and additional services to retain the clientele for services in spheres which are not directly relevant to its main goals."[14]

Finally, the symbolic interactionist tradition has recently been represented by the work of Glaser and Strauss. Their paradigm of "contexts of awareness" is designed to explain the interpersonal contingencies of dying in a hospital. A critical aspect of this scheme is the fact that staff and patients often interact in terms of very different definitions of the situation. They conclude that, ". . . in so much writing about interaction there has been much neglect or incomplete handling of *relationships* (italics ours) between social structure and interaction that we have no fear of placing too much emphasis upon those relationships . . . the course of interaction may partly change the social structure within which interaction occurs."[15]

Suggestive as these remarks are, their essentially descriptive character has precluded a realization of their points of convergence and their analytic potential. The purpose of the remainder of this paper is to set forth a model of formal organizations with two uses: first, to provide a frame of reference which facilitates a synthesis of previous work dealing with clients and organizations; and second, to provide an analytical point of departure from which other hypotheses may be generated concerning relationships between organizations and their clients.

A Perspective toward Clients and Organizations

Notwithstanding the apparently divergent interests in the works referred to above, a major theme is discernible, *viz.*, organizations have contrasting interests in their clients. Furthermore, these organizational interests in the "client biography" may vary along two major dimensions. First, such interests may range from a highly truncated span of time (as in the emergency room of a general hospital) to an almost indeterminate span of time (as in a long-term psychiatric facility or a chronic illness hospital). There is, moreover, a second range of interests which considers the client not in terms of biographical time, but rather in terms of biographical space. That is to say, some organizations may have an interest in only a limited aspect of the client as a person—as in the case of a short-term general hospital—whereas other organizations may have a more extended interest in who the client is as a product of and participant in society—as in the case of a psychiatric out patient clinic.

The analytically important fact is that lateral and longitudinal interests in the biographical careers of clients may vary independently of one another. There are four logically different kinds of arrangement—each of which is likely to have significantly different impacts upon the internal structure and interpersonal processes of organizations, as well as upon extra-organizational relationships. The four biographical variants may be depicted as follows:

	Biographical interest	
Empirical examples	*Lateral (social space)*	*Longitudinal (social time)*
Acute general hospital	–	–
TB hospital, rehabilitation hospital, public health department, medical school	–	+
Short-term therapeutic psychiatric hospital	+	–
Long-term therapeutic hospital, Liberal Arts College	+	+

The logic of this typological system suggests that certain similarities ought to be found between those organizations manifesting a similar lateral interest in their clients, even though they may differ sharply in the extent of their longitudinal concern. Thus, for example, one would expect to find some structural similarities between a general hospital and a tuberculosis hospital, in spite of the fact that the latter has an extended longitudinal interest in the client, while the former does not. That is to say, the orientation of both institutions toward their clients, i.e., patients, is highly specific, focusing as each does upon relatively well-defined disease entities. Thus, though each organization may take account of such lateral life-space factors as occupation, family life, age, and sex, the relevance of these to the defined client problem is minimal. Conversely, those institutions which have a similar stake in the longitudinal careers of their clients should share some features in common despite possible marked differences along the lateral dimension. Thus, a long-term psychiatric hospital, for example, should logically resemble in some respects a tuberculosis hospital, even though the former has a broad-lateral interest in the client, while the latter does not. And similarly, each of the four types should reflect some organizational characteristics which distinguish them.

CLIENT BIOGRAPHIES AND ISSUES OF COMPLIANCE

One of the persisting theoretical issues in organizational analysis has to do with the strategies by which participants are made tractable to the internal needs of organizations.[16] This issue is of equal importance when the client becomes the focus of attention rather than the operative functionaries. The four types of client biographical interests outlined here appear to give rise to different kinds of control problems, and, therefore, to different structural arrangements for achieving compliance.

In utilizing the client as the point of departure by which to examine organizational dynamics, an immediate issue concerns the distinction between conformity and commitment as different modes of client compliance. In the former instance, clients' adherence to conduct rules in the organization is the key problem; in the latter the investment of the client in the ideology of the institution is at issue. These modes of compliance pose different organizational problems in each of the four types. Thus, the greater the laterality of the organization's interest in the client's biography, the greater is the variety of conduct alternatives on the part of the client which are regarded as organizationally relevant. This sets the stage for the emergence of contrasting control strategies. Conversely, those institutions with a minimal lateral interest in their clients are likely to b those in which the conformity of clients to organizational rules is of less concern. In extreme examples, in fact, conformity may be regarded as given and hence unproblematic, because of the physical structure of the institution, e.g., close security cells in custody prisons, or by the physical incapacitation of the client, e.g., quadriplegics in rehabilitation hospitals.

In the longitudinal institution, however, the compliance problem is of a somewhat different order since such organizations have a long-term commitment to the client's future biography which in some cases may extend beyond the time he will actually be physically present in the institution. In these circumstances, the re-arrangement of the client's future biography cannot be accomplished merely by the exercise of coercion. It would appear that for this type of institution the client is controlled by getting him to believe either in the moral goodness or in the practical fitness of the biography the organization is attempting to shape for him. This problem is often attacked by way of an elaborate ideology which the organization attempts to transmit to the client so that self-control is exercised once he is outside the physical confines of the institution.[17]

In terms of the client biography model, the patterns of conformity and commitment take the following shape:

Orientations toward clients		*Compliance problems*	
Lateral	*Longitudinal*	*Conformity*	*Commitment*
–	–	No	No
+	+	Yes	Yes
–	+	No	Yes
+	–	Yes	No

In summary to this point: These problems of client control, which derive logically from a presumed differential institutional investment in the biographies of their clients (laterally and/or longitudinally), give rise to a series of different types of organizational problems and are attended by different modes of resolution.

CLIENT BIOGRAPHIES AND PROBLEMS OF STAFF CONSENSUS

In addition to the contrasting problems of client conformity which derive from the model, the organization's concern with client biography also gives rise to contrasting problems of staff consensus. That is, organizations also may be described by the extent to which conflict between and among staff members is present with regard either to means or to ends. It is our contention that the patterns of consensus relevant to organizational means and ends are systematically related to laterality and longitudinality. Specifically, they take the following form:

Orientations toward clients		*Difficulties over consensus*	
Lateral	*Longitudinal*	*Means*	*Ends*
–	–	No	No
+	+	Yes	Yes
–	+	No	Yes
+	–	Yes	No

With respect to the non-lateral and non-longitudinal institution, the specificity of the orientation toward clients results in clear priorities and consensus as to the relative efficacy of different skills in the repair job to be done. Hence there is little ground for competing orientations to be developed. Similarly, this specificity of orientation and subsequent instant removal of the client implies that there is no compulsion to devise criteria or mechanisms for evaluating long-term outcome, the allocation of organizational resources for these purposes, nor need to establish boundaries to longitudinal responsibility.

This does not mean that stress and strain do not occur in the non-lateral/non-longitudinal organization. It means simply that they seldom become subject to *formal procedures*, but occur at the informal and extra-institutional level. Thus, claims for status are made by those whose place in the hierarchy of professional priorities is somewhere other than the top.[18] Informal negotiations are engaged in for scarce organizational resources. Power alignments develop among staff, involving agreement of a *quid pro quo* kind.[19] Moreover, such an institution is continually subject to pressures from outside, generally in the direction of pressing for greater laterality and longitudinality. Internally as well, informal negotiations develop regarding the ultimate goals and purposes of the instituion, again in the direction of more broadly defining the goals of the establishment.

The most contrasting situation with regard to staff competition and conflict is to be found in the lateral and longitudinal organization, in which there is a heightened organizational response to the ubiquitous pressures for formal resolution which stem from the existence of diverse postures toward means and ends. In view of the felt need for official consensus regarding means and ends, such an organization continually devises officially established devices for making such resolutions. While the initial roots of conflict regarding means and ends may well emerge from within the context of the informal system of power alignments and personal negotiations, these issues are swiftly legitimized and made subject to formal means of solution. Here is to be found a proliferation of formal systems of communication, specialized staff meetings, and increased attempts to make the organization conform to some popularized conception of bureaucracy. The not infrequent outcome is a repeated re-organization of the system of authority and decision-making, and continual addition of staff personnel with finely discriminated skills and techniques. In short, the lateral-longitudinal organization involves a continually changing formal system of authority, with a conflict culture the content of which is co-opted into the formal system.

Although there are other obvious consequences of laterality and longitudinality for the internal structure and dynamics of organizations, we turn now to a consideration of some of their consequences for one type of inter-organizational dilemma, namely, collaborative relations between organizations.

CLIENT BIOGRAPHIES AND INTER-ORGANIZATIONAL COLLABORATION

It is useful to make the distinction between formal and informal processes of inter-organizational collaboration. We shall define formal processes as those ways in which members of organizations engage in collaboration *in their capacities as members of the organization*. By informal we mean those ways of collaborating which involve either an *intervening* organization, e.g., a professional association, or those in which the collaborators act in some capacity other than as organizational members, e.g., a voluntary community organization. Finally, we emphasize the importance of the distinction between administrative-financial concerns as compared with collaboration involving operational facilities. It seems reasonable to argue that these modes of inter-organizational collaboration are also systematically related to the character of the organization's interest in the client's biography. With respect to these distinctions we suggest specifically that the four types of organizations differ in their *propensities* for kinds of collaboration.

The non-lateral/non-longitudinal organization (the acute general hospital, for example), typically has little propensity for formal collaboration

Orientations toward clients		Modes of collaboration			
		Formal		Informal	
Lateral	Longitudinal	Operating	Admin.	Operating	Admin.
–	–	No	No	Yes	Yes
+	+	Yes	Yes	No	No
–	+	Yes	No	No	Yes
+	–	No	Yes	Yes	No

at either the administrative or operating levels. The specificity of its interest in the client and its concern with discriminating strategies of care tend to make such organizations isolated professional islands in the community.[20] Moreover, while this situation may result in the efficient operation of separate institutions, such efficiency does not necessarily extend to the community as a whole. In fact, the reverse may indeed be true—that is, the very efficiency of separate institutions may imply duplication of expensive services such as a cobalt machine, and may thus be detrimental for the needs of the community which they independently serve.[21]

In addition, because they do have a truncated longitudinal interest in their clients, such organizations need not devise strategies to follow their departed clients. "Checking" on clients requires the development of administrative mechanisms for getting information from other organizations which may later be responsible for the welfare of the client. In addition, such organizations normally stand as splendid pillars of financial isolation in the community, with little need (or capacity) to develop "master plans" with other organizations.[22] But again, this is only at the formal level; such organizations are involved in networks of informal relationships. In the case of the general hospital, for example, such networks may extend through the local medical society and health insurance programs in the community as well as to the local community power structure. We do not mean that the non-lateral/non-longitudinal organization does not engage in collaboration, but only that control of the kind and extent of collaboration has been co-opted by extra-organizational agencies.

On the other hand, the lateral-longitudinal organization stands as the most contrasting type. The long-term therapeutically oriented psychiatric institution, for example, is customarily involved in a massive and sometimes conflicting set of administrative and operating linkages at both the formal and the informal level. The wide range of professional personnel it utilizes tends to extend their professional contacts into other similarly organized institutions. Further, the fact of a longitudinal interest in the client's future biography means that the organization must devise ways of establishing working relationships with other organizations which may ultimately be held responsible for the later career of the client. Thus one is likely to find that the non-lateral/non-longitudinal organization (hospital or not) has no established linkages with the juvenile court, nursing homes, family welfare agencies, the probation office, and so forth, while the administrative personnel in the longitudinal institution are often intimately tied in with a wide range of other interested institutions.[23]

In sum, there appear to be variable relations between an organization's structural extensions in time and space toward other organizations, and its functional commitments to the client.

For purposes of this paper, we shall not pursue the other two types. It is rather more strategic on both theoretical and practical grounds to consider what is likely to happen when two organizations of the same type or of sharply divergent type are faced with a potential collaborative relationship. We would expect that a similarity in laterality or longitudinality would be likely to enhance formal collaboration, while contrasting types would be inhibited in collaboration and even experience open conflict.

In the field of rehabilitation, for example, one may find illustrations of these divergent types. Deliberately contrived programs of collaboration involving the consolidation of different rehabilitation agencies, such as organizations for the blind, the mentally retarded, or the physically handicapped, often founder at the operational level. This situation may be explained by the fact that rehabilitation agencies are differentially committed to the lateral careers of their clients. What appears to account for the collaborative effort in the first place is their common interest in the longitudinal dimensions of the client biography. The logical outcome of this duality leads to harmony in terms of effective dialogue at the administrative level but to a great deal of conflict and stress at the operational level. Furthermore, this condition may become characterized over time by elaborate administrative superstructures rather than by operational effectiveness.[24]

These illustrations point to but a few of the logical outcomes for collaboration problems between organizations which stem from the client biography model herein considered. We would expect that the nature of the analysis indicated would also be relevant and useful for an understanding of the organizational dilemmas encountered by such agencies as public health facilities, custodial and punishment-centered institutions, schools, and other client-oriented organizations.

Summary and Conclusion

The client biography model discussed in this paper provides a framework conducive to a more systematic linkage between four major, but often divisive, orientations associated with organizational analysis; namely, the classical bureaucratic, the systemic, the communal, and the symbolic interactionist traditions. The importance of this potential is underscored by the fact that although sociologists are generally aware of the need to better integrate these orientations, attempts to do so have tended to remain implicit and have failed to specify the theoretical link between clients and organizations. This is not to say that the importance of clients in organizations has been overlooked—the point to be emphasized is that existing theories have not incorporated client characteristics in the propositions with which they deal.

The conception of relations between organizations and their clients as varying along the lateral and longitudinal dimensions may be regarded as an initial step toward just such a synthesis.

NOTES

1. Warren Bennis, "Beyond Bureaucracy," *Transaction*, 2 (July–August, 1965), pp. 31–35.

2. See, for example, T. R. Anderson and S. Warkov, "Organizational Size and Functional Complexity: A Study of Administration in Hospitals," *American Sociological Review*, 26 (February, 1961), pp. 23–28; Peter M. Blau, *The Dynamics of Bureaucracy*, Chicago: University of Chicago Press, 1955; Amitai Etzioni, *A Comparative Analysis of Complex Organizations*, New York: The Free Press of Glencoe, 1961; Alvin Gouldner, *Patterns of Industrial Bureaucracy*, Glencoe: The Free Press, 1954.

3. For example Ivan Belknap and J. Steinle, *The Community and Its Hospitals*, Syracuse: Syracuse University Press, 1963; Ray H. Elling, "The Hospital Support Game in Urban Center," in Eliot Freidson, ed., *The Hospital in Modern Society*, Glencoe: The Free Press, 1963; Basil Georgopoulous and F. Mann, *The Community General Hospital*, New York: Macmillan Co., 1962; Delbert Miller, "Industry and Community Power Structure: A Comparative Study of an American and an English City," *American Sociological Review*, 23 (February, 1958), p. 9–15; Harold W. Pfautz and G. Wilder, "The Ecology of a Mental Hospital," *Journal of Health and Human Behavior*, 3 (Summer, 1962), pp. 67–72; Stanley Lieberson, "Ethnic Groups and the Practice of Medicine," *American Sociological Review*, 23 (October, 1958), pp. 542–549.

4. Philip Selznick, "Foundations of the Theory of Organizations," *American Sociological Review*, 13 (February, 1948), pp. 2–35; *TVA and The Grass Roots*, Berkeley: University of California Press, 1953; Talcott Parsons, "Suggestions for a Sociological Approach to the Theory of Organizations," *Administrative Science Quarterly*, 1 (June, 1956), pp. 63–85.

5. For example, J. Bensman and I. Gerver, "Crime and Punishment in the Factory,"

n A. Gouldner and H. Gouldner, eds., *Modern Society*, New York: Harcourt Brace and World, 1963, pp. 593–596; Barney Glaser and Anselm Strauss, *Awareness of Dying*, Chicago: Aldine Press, 1965; Erving Goffman, *The Presentation of Self in Everyday Life*, Edinburgh: University of Edinburgh Press, 1956; Julius Roth, *Timetables*, Indianapolis: Bobbs-Merrill, 1963.

6. For example, Michel Crozier, *The Bureaucratic Phenomenon*, Chicago: University of Chicago Press, 1964; Eugene Haas, R. Hall and N. Johnson, "The Size of Supportive Components in Organizations," *Social Forces*, 42 (October, 1963), pp. 9–17; Robert Merton, "Bureaucratic Structure and Personality," in *Social Theory and Social Structure*, Glencoe: The Free Press, 1949, pp. 151–160; Melvin Seeman and J. Evans, "Stratification and Hospital Care: I. The Performance of the Medical Interne," *American Sociological Review*, 26 (February, 1961), pp. 67–80; Arthur Stinchcombe, "Bureaucratic and Craft Administration of Production," *Administrative Science Quarterly*, 4 (September, 1959), pp. 168–187; Stanley Udy, Jr., "Bureaucratic Elements in Organizations: Some Research Findings," *American Sociological Review*, 23 (August, 1958), pp. 415–418.

7. For example, Blau and Scott *op. cit.*, especially chapter 3, "The Organization and Its Publics," pp. 59–86; Burton R. Clark, *The Open Door College*, New York: McGraw-Hill, 1960; Charles Perrow, "Goals and Power Structures: A Historical Case Study," in Eliot Freidson, ed., *op. cit.*, pp. 112–146; Erwin Smigel, "The Impact of Recruitment on the Organization of the Large Law Firm," *American Sociological Review*, 25 (February, 1960), pp. 56–66; James D. Thompson and W. McEwen, "Organizational Goals and Environment: Goal Setting as an Interaction Process," *American Sociological Review*, 23 (February, 1958), pp. 23–31.

8. The classic discussion of this issue is found in James March and H. Simon, *Organizations*, New York: John Wiley, 1958; a specific statement of the relationship between incentives and organizational types is found in Peter B. Clark and J. Q. Wilson, "Incentive Systems: A Theory of Organizations," *Administrative Science Quarterly*, 6 (September, 1961), pp. 129–166.

9. For example, Fred Davis, "Definitions of Time and Recovery in Paralytic Polio Convalescence," *American Journal of Sociology*, 61 (May, 1956), pp. 582–587; Barney Glaser and Anselm Strauss, "Temporal Aspects of Dying as a Non-Scheduled Status Passage," *American Journal of Sociology*, 71 (July, 1965), pp. 48–59; Erving Goffman, "The Moral Career of the Mental Patient," in *Asylums*, New York: Doubleday, 1961, pp. 125–170; Erving Goffman, *The Presentation of Self in Everyday Life*, Edinburgh: University of Edinburgh Press, 1956.

10. Talcott Parsons, "Suggestions for a Sociological Approach to the Theory of Organizations," in A. Etzioni, ed., *Complex Organizations: A Sociological Reader*, New York: Holt, Rinehart and Winston, 1961, pp. 39–40.

11. Blau and Scott, *op. cit.*, p. 77.

12. Charles Perrow, "Hospitals: Technology, Structure, and Goals," in James March, ed., *Handbook of Organizations*, Chicago: Rand-McNally, 1965, pp. 650–677.

13. Amitai Etzioni, *Modern Organizations*, Englewood Cliffs: Prentice-Hall, 1964, p. 94.

14. S. N. Eisenstadt, "Bureaucracy, Bureaucratization, and Debureaucratization," in Etzioni, *Complex Organizations: A Sociological Reader*, p. 276.

15. Glaser and Strauss, *Awareness of Dying*, *op. cit.*, p. 284.

16. Amitai Etzioni, "Organizational Control Structure," in James March, ed., *Handbook of Organizations*, *op. cit.*, pp. 650–677.

17. A dimension of clients in organizations which is not pursued here has to do with the instrinsic content of the socialization process and its effects upon the individual. A major issue along these lines has to do with the consequences of conformity for

behavior expectations on the one hand, and for the internalization of values on the other. See, for example, Robert Dubin, "Deviant Behavior and Social Structure," *American Sociological Review*, 24 (April, 1959), pp. 147–164; and Irving Rosow, "Forms and Functions of Adult Socialization," *Social Forces*, 44 (September, 1965), pp. 35–45.

18. One of the key organizational issues which stems from lateral interests, particularly in psychiatric institutions, has to do with the presumption of rank-equality among clinical staff. See for example, Milton Greenblatt, R. York and E. Brown, *From Custodial to Therapeutic Patient Care in Psychiatric Hospitals*, New York: Russell Sage Foundation, 1955; Mark Lefton, S. Dinitz and B. Pasamanick, "Decision-Making in a Mental Hospital: Real, Perceived, and Ideal," *American Sociological Review*, 24 (December, 1959), pp. 822–829; Robert Rapoport and Rhona Rapoport, "Democratization and Authority in a Therapeutic Community," *Behavioral Sciences*, 2 (April, 1957), pp. 128–133; William Rosengren, "Communication, Organization, and Conduct in the 'Therapeutic Milieu'," *Administrative Science Quarterly*, 9 (June, 1964), pp. 70–90.

19. For example, Richard McCleery, "Authoritarianism and the Belief System of Incorrigibles," in D. Cressey, ed., *The Prison: Studies in Institutional Organization and Change*, New York: Holt, Rinehart and Winston, 1961, pp. 260–306; William R. Rosengren and S. DeVault, "The Sociology of Time and Space in an Obstetrical Hospital," in Freidson, *op. cit.*, pp. 266–292; Anselm Strauss, *et al.*, "The Hospital and Its Negotiated Order," in E. Freidson, ed., *op. cit.*, pp. 147–169.

20. See, for example, Ray Elling, "The Hospital Support Game in Urban Center," in Freidson, *op. cit.*, pp. 73–112; Oswald Hall, "The Informal Organization of the Medical Profession," *Canadian Journal of Economics and Political Science*, 12 (February, 1946), pp. 30–44.

21. For example, J. H. Robb, "Family Structure and Agency Co-ordination: Decentralization and the Citizen," in Mayer N. Zald, *Social Welfare Institutions: A Sociological Reader*, New York: John Wiley, 1965, pp. 383–399; Oliver Williams, *et al.* *Suburban Differences and Metropolitan Policies: A Philadelphia Story*, Philadelphia: University of Pennsylvania Press, 1965.

22. See, for example, Charles V. Willie and Herbert Notkin, "Community Organization for Health: A Case Study," in E. Gartley Jaco, *Physicians, Patients, and Illness*, Glencoe: The Free Press, 1958, pp. 148–159.

23. For example, Sol Levine and P. White, "Exchange as a Conceptual Framework for the Study of Interorganizational Relationships," *Administrative Science Quarterly*, (March, 1961), pp. 583–601; Eugene Litwak and L. Hylton, "Interorganizational Analysis," *Administrative Science Quarterly*, 6 (March, 1962), pp. 395–420; J. V. D. Saunders, "Characteristics of Hospitals and of Hospital Administrators Associated with Hospital-Community Relations in Mississippi," *Rural Sociology*, 25 (June, 1960), pp. 229–23; James D. Thompson, "Organizations and Output Transactions," *American Journal of Sociology*, 68 (November, 1962), pp. 309–324.

24. A clear example of this process can be discerned in the recent history of the National Mental Health Association. A short time ago this existed merely as a loosely held together congeries of autonomous, local mental health societies. Some of these groups were laterally and others non-laterally committed to their clientele. However, they shared in common a longitudinal interest in the careers of locally-defined client groups. The original move toward official collaboration came through the New York office and has persisted at the administrative and fund-raising level. It has now reached the point where most of the originally autonomous local societies provide little or no service to clients. They function merely as linkages in a nationwide administrative system. It should be added, lastly, that this decline of service functions and pre-eminence of administrative functions had also resulted in a dramatic shift in the sources of recruitment and staffing patterns of these organizations.

PART III

The Impact of Internal Factors on Organizations

The basic theme of this section is identical to that of the preceding one—the several internal factors which have been identified as having an impact on organizations will be discussed and the direction of their impact suggested. This similarity in theme will be accompanied by a similarity in conclusion. That is, it is not yet possible to specify the extent to which each factor to be identified affects organizations in relation to the other factors discussed. We do not know, for example, if the presence of professionals in organizations has a greater or lesser impact on the organization than the forms of leadership which are exercised. As in the case of the external factors, insufficient attention has been paid to comparing the impact of the several internal relationships which have been intensively studied.

Today, no scholar or practitioner interested in organizations ignores the importance of internal developments for the understanding of organizations. Data have been presented which overwhelmingly indicate that the members of organizations have a direct bearing on what actually goes on in the organization. This bearing is usually viewed as being some kind of deviation from the officially prescribed arrangements. There have been extensive arguments regarding the functionality of dysfunctionality of such deviations. Fortunately, these arguments have subsided and in their place have arisen inquiries regarding the conditions under which particular forms of deviations occur and the consequences of the deviations for the individuals and organizations involved. The assumption is no longer made that these deviations are unusual or good or bad. The intent is now to specify the conditions under which such deviations arise and to trace their outcomes. Interestingly, there have been few attempts to look at deviance in organizations by students of deviant behavior, even though organizations are a remarkably easier research setting than the wider community.

The early identification of these forms of deviation led to the development of terms such as "informal" or "unofficial" organizations in contrast to the "formal" organization as it is established by the organization itself. This is an analytical distinction which actually does not have practical or theoretical benefits. What the major concern of organizational analysts is and should be is the behavior of individuals and the organization itself. This behavior is a combination of the officially prescribed system and the derivations developed through social interactions. There is, therefore, only an analytical distinction between formal and informal systems, with actual behavior being the crucial element for research and practice.

The recognition that organizational practice did not always follow organizational prescription contributed to an intensive interest in the varieties of leadership and/or supervisory patterns which could be utilized in the organizational setting. The issue of leadership is an important one in that the behavior of leaders has an impact on the led and thus on the organization as a whole. (This assumes a reciprocal interaction between the leaders and the led. These comments can also be extended to the broader issues of power relationships.) The behavior of individuals, particularly those in higher positions has a distinct impact on the organization. One of the major anomalies in the organizational literature is in this area. There is a startling absence of research into the impact of the top leaders on the organization with other possibly confounding factors held constant. The management succession literature is inconclusive in terms of the impact that a change in top management makes. It appears that the more tightly structured the organization, the less the impact of such succession. This is an area, however, which requires much more research before definitive conclusions are reached.[1]

The discussion thus far has focused on relationships among organizational members and between superiors and subordinates. A major consideration in terms of the internal factors which affect organizations is the nature of the people who are in the relationships discussed. Research in this area has been developed from several perspectives. At one time there was a great deal of concern about the traits of the entering personnel, especially those in superordinate positions. Emphasis in this area has shifted away from the trait approach, which did not prove to be empirically relevant to a concentration on the situations in which specific forms of leadership emerge. At the same time, other traits of individual organizational members which would appear to have significance for the organization have been surprisingly ignored. There have been relatively few examinations of the role of minority group members (here, of course, traits are ascribed to the individuals), such as Blacks, women, or Indians, on the organization and reciprocally, of the impact of organizational life on the minority group members. Changed distributions of dominant-minority group members in an organization would almost certainly affect the internal dynamics of the organization. An equally important point here is that such demographic changes are usually linked to external conditions, such as general demographic shifts or altered legal requirements. This again illustrates the interplay between internal and external factors in organizations.

One area where the nature of the personnel involved has received a great deal of attention is that in which the personnel bring into the organization their own set of work relevant norms. Professionals, and to some degree craftsmen, have been the subject of intensive investigations as the relative importance of organizational versus personal (professional-craft) norms becomes exceedingly important for the organization. Although the evidence is not yet overwhelming in this area, it appears that organizations are able to accommodate the presence of externally derived norms brought into the organization through members by altering their normative system in those areas in which craftsmen or professionals are employed. Unfortunately, the concentration on professionals has been associated with an absence of concern about other occupational groups which might also bring norms and values with them to the organization and which would in turn affect the organization. The professions and crafts do, however, appear to have enough unique characteristics so that the attention paid to them has been fruitful for our knowledge about organizations.

A major factor which shapes organizations is the relationships between segments of the organization. Professionals and administrators can conflict over evaluation standards. If the professionals succeed in having different evaluative criteria applied to them, the organizational structure is altered. The long history of labor-management conflict and the ensuing changes which have taken place in work rules is another example of internally derived organizational change. Student militancy in schools and colleges has led to important shifts here.

Before turning to the articles contained in this section, an additional internal element deserves comment. This element is the history or traditions of organizations. It is an element which has received almost no attention by sociologists. Stinchcombe has pointed out that the historical eras in which particular types of organizations emerge have important ramifications for their future life, an external influence, but internal historical factors have been almost totally ignored.[2] An equal amount and kind of Black militancy, for example, would probably have different impacts on organizations with histories of open hiring and promotion than on those which seldom hired or promoted Blacks. Unfortunately, we do not know what these differences would be.

The articles in this section have been selected to demonstrate the points made above. As is always the case in such a selection process, other papers could have been selected for the same purpose. The selection process here was based around a desire for diversity in perspective. Despite this diversity, the central points of this section have become generally accepted in the organizational literature. There will be no papers dealing with the deviations from the formal normative system by organizational members. This point has been documented extremely well by this time. We can accept as fact that such deviations occur and form different organizational structures than those formally prescribed.

The paper dealing with leadership or supervision attempts to specify the conditions under which leadership will have the greatest influence on the performance of organizational members and thus on the organization. The Stinchcombe and Harris article points out that supervision *per se* will vary in importance according to the conditions in the organization. Thus supervision, regardless of the style used, is not as important in some situations as in others.[3]

From these ideas, which do not span the entire breadth of the controversies surrounding leadership and supervision, it should be evident that this is not a closed subject as some of the more prescriptive writers suggest. Rather, leadership is situationally based, linked to the characteristics of those in the relationship, and relevant only under particular circumstances. At the same time, it should be equally clear, and this is the point of the argument, that at least under some conditions the extent and form of supervisory or leadership behavior makes a difference for the organization.

The next paper in this section deals directly with the interactions between professional and organizational control structures. Hall concludes that organizations and organizational units staffed by professionals are less bureaucratically structured because of the presence of the professionals. For the purposes of this discussion, the central point is that the nature of the personnel in an organization does make a vital difference for the form which the organization takes.

Brager's paper concludes this section. It shows some of the sources of conflicts and the manner in which they emerge in organizations. While the effects on the organization are not discussed, it is clear that differing commitments are an important potential source of intra-organizational conflict. This, together with emerging and changing alliances among members, have important impacts on the organizations involved.

Ideally, the factors which have been identified should be sorted out and their relative strengths under particular conditions specified. Since the field is not yet at the point where this is possible, an awareness of the issues involved and preliminary research into these questions is probably the point from which current departures should be made. In addition, the identification of other such internal factors should proceed.

NOTES

1. Alvin W. Gouldner, *Patterns of Industrial Bureaucracy*, Glencoe: Free Press, 1954; Robert H. Guest, "Managerial Succession in Complex Organizations," *American Journal of Sociology*, 68-1 (July, 1962) pp. 47–54; Oscar Grusky, "Organizational Size and Managerial Succession," *American Journal of Sociology*, 67–3 (November, 1961), pp. 261–269; Louis Kriesberg, "Careers, Organizational Size, and Succession," *American Journal of Sociology* 68–3 (November, 1962), pp. 355–59; Oscar Grusky, "Managerial Succession and Organizational Effectiveness," *American Journal of Sociology*, 69–1 (July, 1963), p. 21; Oscar Grusky, "The Effects of Succession: a Comparative Study of Military and Business Organizations," in Oscar Grusky and George A. Miller (eds.) *The Sociology of Organizations: Basic Studies*, New York, The Free Press, 1970, pp. 439–454.

2. Arthur L. Stinchcombe, "Social Structure and Organizations," in James G. March (ed.) *Handbook of Organizations*, Chicago, Rand McNally and Co., 1965.

3. Amitai Etzioni, in "Dual Leadership in Complex Organizations," *American Sociological Review*, 30–5 (October, 1965), pp. 688–698, takes this argument a step further. He argues that the specific style of leadership which is exercised in organizations should be linked to the goals of the organization. A major implication is that attempts to utilize an inappropriate form of leadership will be counter-productive for the organizations and individuals involved.

6

ARTHUR L. STINCHCOMBE
AND T. ROBERT HARRIS

INTERDEPENDENCE AND INEQUALITY: A SPECIFICATION OF THE DAVIS-MOORE THEORY

The Davis-Moore theory of social stratification (1945) argues that inequality is due to differences in the "importance" of social roles, and that recruitment to important social roles will be more oriented toward getting adequate performance than will recruitment to other roles. Importance of roles, then, tends to create inequality and to create a force toward recruitment by ability and training. One of the difficulties in working with this theory empirically is to obtain independent measures of the importance of social roles.

In this paper we will treat the importance of supervision in the productivity of groups. We will show that for groups doing highly interdependent tasks, the marginal productivity of added amounts of supervision (starting from any given level of supervision) will be higher than in groups doing independent tasks. Further we will show that the ability of a supervisor is more important in determining group productivity in the interdependent case.

The derivation of these results will be purely theoretical. If actual stratification systems (in small groups, factories, or societies) actually behave as the Davis-Moore theory implies, then the machinery here will help derive predictions about the relation of task structure to stratification structure. This involves defining the importance of a social role by how much difference its performance or non-performance makes to total group

Reprinted from Arthur L. Stinchombe and T. Robert Harris, "Interdependence and Inequality: A Specification of the Davis-Moore Theory," *Sociometry*, 32–1 (March, 1969), pp. 13–23.

performance, and by how much difference it makes to group productivity whether it is well performed or not. We do this by applying the techniques of marginal analysis from economics to a mathematical formulation of what it means for tasks to be interdependent.

An Intuitive Introduction to the Argument

Consider the hot processing and cold processing parts of a steel rolling mill. The hot processing part shapes hot steel by passing it through consecutive shaping machines before it gets cold. The cold processing part cuts, grinds, classifies, and otherwise finishes the rolled product.[1] One cannot build up inventories of hot metal between the shaping machines, because maintaining the temperature would require installation of very expensive furnaces and add an input and output worker for each of those furnaces. Cold steel products do not change their state if inventoried temporarily between machines. Under these conditions, the hot shaping machines can only work when the other hot shaping machines on each side of them are functioning. But when a cold processing machine stops, the succeeding machine can work from stock previously produced, and the preceding machine can produce for stock to be processed later by the stopped machine.

Now let us suppose that a single machine breaks down in both sections of the line. A supervisor can, let us suppose, choose where to direct his energies in trying to get the stopped machine running. Let us further suppose that he can get each of the machines running by spending half an hour of time. If he spends a half hour on the hot shaping machine, he will have succeeded in getting the whole hot line running. If he were to spend that half hour on the machine on the cold line, he would get only that machine running, since the others could continue to run anyway. That is, the productivity of his getting a particular machine running on the hot line will be multiplied by all the other machines he gets running. The productivity of getting the machine on the cold line running will be only the productivity of that machine. Hence if he has to choose which to pay attention to first, he will choose to pay attention to the hot line and leave the cold line for later.

Likewise consider the general superintendent's problem in the mill, when deciding where to put his most able supervisor. Suppose, for instance, he has a supervisor whose technical competence is such that he only needs three minutes to diagnose and remedy the problem on either machine, rather than 30 minutes. Clearly if he puts the three minute supervisor on the hot line, he will save 27 minutes of the production of the whole line. If he puts the three minute supervisor on the cold line, he will save only 27 minutes of the production of a single machine. The productivity of ability in supervision is apparently much greater in the case of interdependence than it is in the case of relatively independent production processes.

A Formalization of Interdependence

For simplicity in presentation, let us consider a productive process in which each machine ("role" or any other concrete component of a group process can be substituted for "machine") is either producing or not producing. When it is producing, it produces at a constant rate. Then production of an individual machine is proportional to the probability that the machine will be running at any given time.[2]

First we consider the total production, T_1, when the production of each of the activities is independent; that is, when variations in the production of one machine have no effect on the output of other machines. This is a rough approximation to the short-run situation on the cold line described above. In this case we merely sum the productions of the individual workers:

$$T_1 = \sum_{i=1}^{n} bp_i = nb\overline{p}. \quad (1)$$

where p_i is a standardized measure of the production of the i'th worker (for example, the probability that he will be working at a given time), b relates this measure of productivity to his contribution to total production, p is the arithmetic mean of the numbers $p_1, \ldots, p_n$, and n is the number of workers.

In the case of interdependence as described on the hot line above, the probability that the system will be producing at any given time is the probability that everyone will be working, and hence (assuming that "breakdowns," or failures to produce, are independent for different machines) the average production per unit time is proportional to the product of the probabilities:

$$T_2 = K \prod_{i=1}^{n} p_i \quad (2)$$

where K is the rate of production when the system is producing, and n is the number of machines.[3]

Marginal Productivity of Supervision in Independent and Interdependent Production

Let us compare the marginal productivity of supervision in these two cases. In general, each worker's productivity will depend on the amount of supervision per worker and the ability of supervisors, which we denote by c and a respectively. The marginal productivity of quantity of supervision is the partial derivative of total production with respect to quantity of supervision. In the independent case, equation (1) yields:

$$\frac{\partial T_1}{\partial c} = nb \frac{\partial \bar{p}}{\partial c} = \frac{T_1}{p} \frac{\partial \bar{p}}{\partial c}. \qquad (3)$$

In the interdependent case,

$$\frac{\partial T_2}{\partial c} = K \frac{\partial}{\partial c} \left(\prod_{j=1}^{n} p_j \right) \qquad \text{from equation (2)}$$

$$= K \sum_{i=1}^{n} \left[\left(\prod_{\substack{j=1 \\ j=i}}^{n} p_j \right) \frac{\partial p_i}{\partial c} \right]$$

$$= K \left(\prod_{j=1}^{n} p_j \right) \sum_{i=1}^{n} \frac{1}{p_i} \frac{\partial p_i}{\partial c}$$

$$= T_2 \sum_{i=1}^{n} \frac{1}{p_i} \frac{\partial p_i}{\partial c} \qquad \text{from equation (2)}$$

$$\geqq \frac{T_2}{p_{max}} \sum_{i=1}^{n} \frac{\partial p_i}{\partial c}$$

$$= \frac{nT_2}{p_{max}} \frac{\partial \bar{p}}{\partial c}.$$

Thus

$$\frac{\partial T_2}{\partial c} \geqq \frac{nT_2}{p_{max}} \frac{\partial \bar{p}}{\partial c} \qquad (4)$$

where p_{max} is the largest of $p_1, \ldots, p_n$.

If $T_2 \geqq T_1$, and the characteristics of the workers and their jobs are identically distributed (drawn from the same population) so that for each level a of ability and c of quantity of supervision, p and $\frac{\partial \bar{p}}{\partial c}$ are the same in the two cases,[4] it follows that:

$$\frac{\partial T_2}{\partial c} \geqq n \frac{\bar{p}}{p_{max}} \frac{\partial T_1}{\partial c} \tag{5}$$

where all the symbols are as defined earlier. If we assume also that workers are fairly homogeneous in their productivity,[5] we can conclude from (5) that the marginal productivity of supervision is almost n times as great in the interdependent case as in the independent case (other things being equal).[6]

We can find the marginal productivity of a, the average ability level of supervisors, by differentiating total production with respect to a. The procedure is exactly the same as that above, substituting a for c. The result is:

$$\frac{\partial T_2}{\partial a} \geqq n \frac{\bar{p}}{p_{max}} \frac{\partial T_1}{\partial a}. \tag{6}$$

Thus (other things being equal) a small increase Δa in the level of ability of supervisors will increase production almost n times as much in the interdependent case as would the same increment in the independent case.

Finally, if n is greater than 2 or 3, it is clear that even if our assumptions are only the crudest approximation to reality, the marginal productivity of supervision will still be much greater in the interdependent case than in the independent case.

Empirical Implications of the Argument

Here we will try to state the logic of several implications of the formal argument above, so that they can be applied to a wide variety of cases and give rise to a rich set of empirical consequences. The implications of such a formal argument depend on adding sociological and economic postulates to the analysis of the productivity of supervision. The examples suggested below elaborate the argument in terms of empirically measurable variables.

1. When the marginal productivity of supervision is higher, more supervision will be used. In particular, the more interdependent the activities of a group:

a. the higher the ratio of time spent in supervision to time spent in operative work;
b. the more likely is it that full-time supervisory roles will be established, rather than combinations of supervision with operative duties;
c. together a and b imply that the higher will be the ratio of full-time supervisors to subordinates (the span of control will be smaller if all "supervisors" are formal superiors);
d. hence, in an organization of a given size, the greater will be the number of levels in the hierarchy;
e. If Simon's postulate of a constant ratio of income between hierarchial levels is true, the more inequality there should be in the overall income distribution.[7]

2. When the marginal productivity of supervisory ability is higher, recruitment should tend to be more on the basis of talent, and talent is more likely to be paid a premium.[8] Hence in interdependent processes we should find:

a. more recruitment of supervisors on the basis of qualifications and talent, less seniority or nepotism;
b. more firing, demotion, or transfer of supervisors whose abilities are not adequate to supervisory positions;
c. more differential wages within the group of supervisors of interdependent processes as compared to the variance within the group of supervisors of independent processes;
d. more inequality in the amount of overtime by able versus less able supervisors;
e. more inequality in the informal respect and recognition of ability of supervisors.

3. When rules do not adequately reflect the contingencies of the productive process, we may postulate that informal adaptations to the reality of the situation will take place more rapidly, the more difference they make to group productivity. Informal adaptations to contingencies are a particular social reflection of supervisory ability. Hence we should expect to find in interdependent processes:

a. more communication outside official channels, backed by more urgent emotional force; in particular, we expect more flow upward of communications about problems from subordinates (cf. Blau, 1957: 59–61);
b. more departure of informal, ability-based stratification from formal,

especially formal ascriptive, stratification; in particular, the informal status of competent older line foremen, relative to ascriptively higher status but practically less effective new engineers, should tend to be higher in interdependent processes;

c. more departure of actual activities of different supervisors from official manning charts and job descriptions;

d. more esteem, comparatively, of instrumental leaders as opposed to expressive leaders of work groups.

A Further Line of Development

We have concentrated above on the productivity of the supervisor's role. There is, of course, no inherent reason for this. Any role in the group has a productivity for group performance. If we differentiate the expressions for total production above with respect to the abilities or performances of each worker, something quite similar is derived: the group is more dependent on each worker in the interdependent case.[9]

What sorts of things might follow from this? Perhaps the frequency of attempts to control the behavior of others would increase, and hence more total interaction would ensue. This might provide a metric for Homans' "external system," (1950:90–94), i.e. the amount of externally induced interaction in a group. Hence we would expect to find more solidarity or mutual liking in interdependent groups, because their activities would produce more interaction.

Conversely any deviant can endanger the payoff to everyone in an interdependent group. Consequently we might expect more instability in such groups, more deaths of groups because of individual variability in performance. We might find more such groups ejecting such deviants. We could thus ask the following empirical questions: do more people get fired from the interdependent parts of production lines? Do more businesses fail in industries with interdependent technologies, controlling for size? Are workers more likely to support firings on the interdependent parts of lines? Is rejection of a deviant more enthusiastic, or quicker, in experimental groups with interdependent tasks?

It should be pointed out here that the ratio of rewards of supervisors to the rewards of workers are not predicted by this argument. Both worker ability and supervisory ability are more important in interdependent processes. In fact, if we form the ratios between the marginal productivity of supervisory ability and worker ability, we find for the independent case

$$\frac{\dfrac{\partial T_1}{\partial a}}{\dfrac{\partial T_1}{\partial x_j}} = nR \tag{7}$$

where R is the ratio between the marginal improvement of performance of the average worker with an improvement of supervisory ability and the marginal improvement of a worker's performance with his own increased ability, i.e. $\dfrac{\partial \bar{p}}{\partial a} \Big/ \dfrac{\partial p_j}{\partial x_j}$. The quantity R might be very important in other applications of the Davis-Moore stratification theory, but we have no ideas about the causes of its variation. For the interdependent case, the ratio of marginal productivities of abilities is

$$\frac{\dfrac{\partial T_2}{\partial a}}{\dfrac{\partial T_2}{\partial x_j}} \geqq \frac{p_j}{p_{max}} nR. \tag{8}$$

Except for the term p_j/p_{max}, which will be less than or equal to one, the expressions of the right are identical. If workers are fairly homogeneous in their productivity, p_j/p_{max} will be near to one and the inequality, from equation (4), will be close to an equality. Then we would expect the ratios of supervisory to worker rewards for ability to be about the same in the two cases. The greater importance of supervision in the interdependent case does *not* imply greater *relative* importance of supervisors compared to workers.

Summary and Conclusions

The purpose of this paper was to give an analysis of the "importance" of a social role and to show how it varied under different conditions. We chose in particular to analyze the importance of supervision under conditions of independent activity and under conditions of interdependent activity. We showed that at a given level of supervision, either an added quantity, or an increase of ability, of supervision had more effect when the upervised activities were interdependent. This argument depends on the ccuracy of our formalization of the meaning of interdependence. From his it should follow from the Davis-Moore functional theory of stratifica- ion that (1) roles of supervision are more likely to be differentiated from ubordinate roles, and to be more numerous, in more interdependent ctivities, (2) recruitment by ability and differential rewards for able upervision will be more common in interdependent systems, and (3) nformal adaptation to contingencies will be more common in interde- endent systems.

These results should apply to all kinds of groups to which the Davis- Moore theory is supposed to apply. But groups in factories or other atural settings are almost always embedded in larger stratification sys- ems, created by collective contracts, by labor law, by kinship pressures, y pressure toward equilibrium wages in labor markets, and so forth. urther, we have defined "supervision" abstractly as any activity which mproves the performance of a worker. Not all the supervisors in this ense are in formal authority over workers in natural settings. Especially in nodern industrial practice, staff people do much of what we have called upervision. Natural groups probably depart radically from our predictions ecause of the embedding of the task group in larger social structures.

But tasks of groups without previous internal differentiation are easily nanipulated in social psychology laboratories. We have argued above hat a rich and various body of hypotheses can be derived about stratifica- ion patterns in groups with varying tasks. The way seems open for xperimental exploration of one of the classical problems of the theory of ociety.

NOTES

1. The research in a steel plant from which this example comes was done by Stinch- ombe, under a grant from the Olivetti Foundation to the Joint Center for Urban Studies f MIT and Harvard.

2. The following derivation is not restricted to this case, but is valid whenever the roduction of individuals, however conceived, is related to total production in the same ay as we indicate below.

3. We assume the two groups compared have the same number of workers, in order simplify matters. To avoid confounding the effects of size with those of interdepend- ce on the value of supervision, we hold the former constant. The comparative impact f growth in size on the supervisory problems of independent and interdependent groups ould be explored with the theoretical technology developed below, by taking differences r successive values of n, after differentiating with respect to a or c.

4. These are reasonable restrictions. Since our purpose is to study the effects of inter-

dependence or independence, it is appropriate to "control" other factors. If, for example the nature of individual jobs were different, resulting in a different effect of supervision on the production of individual workers $\left(\frac{\partial \bar{p}}{\partial c}\right)$, then this fact as well as interdependence would affect the marginal productivity of supervision on the total output of the group But this is irrelevant to our argument, which is that the same effects on components have a much greater impact on the total output of the system in the case of interdependence

5. It is possible to assume instead that $\frac{\partial p_i}{\partial c}$ is uncorrelated with $1/p_i$. This assump-tion is met, in particular, if $\frac{\partial p_i}{\partial c}$ is the same for all values of p_i. We can also relax an assumption implicit in equation (1) by rewriting that equation:

$$T_1 = \sum_{i=1}^{n} b_i p_i \tag{1a}$$

where b_i, the contribution of worker i to total production, may now vary with i. How-ever, we must add the assumptions that p_i and $\frac{\partial p_i}{\partial c}$ are uncorrelated with b_i. Under these assumptions, we can derive the marginal productivity of supervision in the inter-dependent case, beginning with the fourth line of the derivation of equation (4):

$$\frac{\partial T_2}{\partial c} = T_2 \sum_{i=1}^{n} \frac{1}{p_i} \frac{\partial pi}{\partial c}$$

$$= T_3 \left(\sum_{i=1}^{n} \frac{1}{p_i} \right) \left(\frac{1}{n} \sum_{i=1}^{n} \frac{\partial p_i}{\partial c} \right) \quad \text{because } \frac{\partial p_i}{\partial c} \text{ is uncorrelated with } 1/p_i$$

$$= \frac{T_2 n}{\tilde{p}} \frac{\partial \bar{p}}{\partial c}$$

where $\tilde{p}$ is the harmonic mean of $p_1, \ldots, p_n$, that is,

$$\frac{1}{\tilde{p}} = \frac{1}{n} \sum_{i=1}^{n} \frac{1}{p_i} .$$

It is well known that $\tilde{p} \leq \bar{p}$, and that the equality holds if and only if $p_1 = p_2 = \ldots = p$ From equation (1) for independence,

$$T_2 = \sum_{i=1}^{n} b_i p_i = \left(\sum_{i=1}^{n} b_i \right) \bar{p} \quad \text{because } b_i \text{ and } p_i \text{ are assumed uncorrelated.}$$

Thus

$$\frac{\partial T_1}{\partial c} = \left(\sum_{i=1}^{n} b_i \right) \frac{\partial \bar{p}}{\partial c}$$

$$= \frac{T_1}{\bar{p}} \frac{\partial \bar{p}}{\partial c}.$$

ence of $T_2 \geqq T_1$, then

$$\frac{\partial T_2}{\partial c} = \frac{T_2 n}{\bar{p}} \frac{\partial p}{\partial c} \geqq \frac{T_1 n}{\bar{p}} \frac{\partial p}{\partial c} = n \frac{\partial T_1}{\partial c}.$$

his result is slightly stronger than (5). Similar remarks apply to inequality (6) below.

6. This is the result that was suggested in the intuitive introduction to the argument, nd it follows for the same reason. The usual mathematical proof of the rule for the erivative of a product of functions, used in deriving the inequality (4), is essentially a eneralization of the intuitive idea presented in the introduction.

7. Herbert Simon (1957). Simon is here following the suggestion by Roberts (1956: 35n). Strictly speaking, this argument depends on there being a single income rate at ach level. As we will show later, interdependence ought also to increase the variance f productivity of operative people at different levels of ability, and it may be that this ariance within levels interacts with the variance between levels in such a way as to lower e overall variance.

8. Compare Stinchcombe's earlier (1963: 806–807) somewhat vague suggestions bout "talent complementary" and "talent additive" industries and activities. The result bove may be restated as: "in interdependent acivities, supervisory talent is more of a omplementary factor of production, than in independent activities."

9. If we differentiate the total production with respect to the performance of the jth orker, p_j, equation (1) gives:

$$\frac{\partial T_1}{\partial p_j} = b = \frac{T_1}{n\bar{p}}$$

r the case of independence. From equation (2) we obtain for the interdependent case:

$$\frac{\partial T_2}{\partial p_j} = K \prod_{\substack{i=1 \\ i \neq j}}^{n} p_i = \frac{T_2}{p_j}.$$

hus the marginal productivity of the jth worker's performance is about n times as great the interdependent case, but varies according to his absolute level of performance. ote that the marginal productivity is greatest for those whose total performance is nallest (p_j then being small). Those with lowest productivity are disproportionately armful.

Suppose that we conceive of a worker's performance as depending on his own ability s well as on the amount of supervision he receives. Let us call the jth worker's ability x_j. hen, by the chain rule for differentiation, applied to the results in the preceding para- raph,

$$\frac{\partial T_1}{\partial x_j} = \frac{T_1}{n\bar{p}} \frac{\partial p_j}{\partial x_j}, \text{ and}$$

$$\frac{\partial T_2}{\partial x_j} = \frac{T_2}{p_j} \frac{\partial p_j}{\partial x_j}.$$

hus the organization stands to gain about n times as much in the interdependent case y raising the workers' level of ability.

REFERENCES

lau, Peter
1957 "Formal organization: dimensions of analysis." American Journal of Sociology 63 (July): 58–69.

Davis, Kingsley and Wilbert E. Moore
1945 "Some principles of stratification." American Sociological Review 10 (April): 242–249.
Homans, George C.
1950 The Human Group. New York: Harcourt, Brace and World.
Roberts, David R.
1956 "A general theory of executive compensation based on statistically teste propositions." Quarterly Journal of Economics 70 (May): 270–294.
1957 "The compensation of executives." Sociometry 20 (March): 32–35.
Stinchcombe, Arthur L.
1963 "Some empirical consequences of the Davis-Moore theory of stratification." American Sociological Review 28 (October): 805–808.

ICHARD H. HALL

ROFESSIONALIZATION ND BUREAUCRATIZATION*

wo related but often non-complementary phenomena are affecting the ocial structure of Western societies today. The first of these is the ncreasing professionalization of the labor force. Occupational groups that ave held the status of "marginal professions" are intensifying their fforts to be acknowledged as full-fledged professions. Occupations that ave emerged rather recently, and some that have not previously been ought of as professions, are also attempting to professionalize. At the ame time, work in general is increasingly becoming organizationally ased. This is true among both the established professions and the profes- onalizing occupations. The intent of this paper is to examine the profes- onalization process in the context of the organizational structures in hich professional or professionalizing workers are found, in order to etermine how these phenomena affect and are affected by each other. ata from a variety of occupational groups found in a variety of organiza- onal settings will be used in this analysis.

Background

iscussions about the nature of professions typically revolve around the rofessional model. The professional model consists of a series of attributes hich are important in distinguishing professions from other occupations.

eprinted from Richard H. Hall, "Professionalization and Bureaucratization," *Ameri- n Sociological Review*, 33–1, (February, 1968), pp. 92–104.

* Grateful acknowledgment is given to Grant GS 882 from the National Science oundation which provided support for the project from which this report is taken.

Movement toward correspondence with the professional model is th process of professionalization.[1] The attributes of the model are of tw basic types. First are those characteristics which are part of the structu of the occupation, including such things as formal educational an entrance requirements. The second aspect is attitudinal, including th sense of calling of the person to the field and the extent to which he us colleagues as his major work reference.

The structural side of the professional model has been intensivel examined by Wilensky, who noted that occupations pass through a rath consistent sequence of stages on their way to becoming professions Wilensky includes the following attributes in his discussion:

1. Creation of a full time occupation—this involves the performance c functions which may have been performed previously, as well as new fun tions, and can be viewed as a reaction to needs in the social structure.

2. The establishment of a training school—this reflects both the know edge base of a profession and the efforts of early leaders to improve th lot of the occupation. In the more established professions, the move is the followed by affiliation of the training school with established universitie In the newer professions, university affiliation is concurrent with the esta lishment of training schools.

3. Formation of professional associations—the formation of such ass ciations often is accompanied by a change in the occupational title, a tempts to define more clearly the exact nature of the professional task and efforts to eliminate practitioners who are deemed incompetent by th emergent professionals. Local associations unite into national associatio after a period of some political manipulations. As stronger associatio are formed, political agitation in the form of attempts to secure licensi laws and protection from competing occupations becomes an importa function.

4. Formation of a code of ethics—these ethical codes are concerne with both internal (colleague) and external (clients and public) relation They are designed to be enforced by the professional associations the selves and, ideally, are given legal support.[3]

A professional attribute that is both structural and attitudinal is th presence of professional autonomy.[4] While the structural aspect of auto omy is indirectly subsumed under the efforts of professional associatio to exclude the unqualified and to provide for the legal right to practic utonomy is also part of the work setting wherein the professional is xpected to utilize his judgment and will expect that only other profesionals will be competent to question this judgment. The autonomy ttribute also contains an attitudinal dimension: the belief of the profesional that he is free to exercise this type of judgment and decision naking.

The attitudinal attributes of professionalism reflect the manner in which he practitioners view their work. The assumption here is that there is ome correspondence between attitudes and behavior. If this assumption s correct, then the attitudes comprise an important part of the work of he professional. If he or his occupation has met the structural prerequisites of professionalism, the approach taken in practice becomes the mportant consideration. The attitudinal attributes to be considered here re:

1. The use of the professional organization as a major reference—this nvolves both the formal organization and informal colleague groupings as he major source of ideas and judgments for the professional in his work.[5]
2. A belief in service to the public—this component includes the idea of ndispensability of the profession and the view that the work performed enefits both the public and the practitioner.[6]
3. Belief in self regulation—this involves the belief that the person best ualified to judge the work of a professional is a fellow professional, and he view that such a practice is desirable and practical. It is a belief in olleague control.[7]
4. A sense of calling to the field—this reflects the dedication of the proessional to his work and the feeling that he would probably want to do he work even if fewer extrinsic rewards were available.[8]
5. Autonomy—this involves the feeling that the practitioner ought to be ble to make his own decisions without external pressures from clients, hose who are not members of his profession, or from his employing rganization.[9]

The combination of the structural and the attitudinal aspects serves as he basis for the professional model. It is generally assumed that both spects are present to a great degree in highly professionalized occupaions, while they are present to lesser degrees in the less professionalized ccupations. Whether or not this is the case will be examined in this esearch. It may be, for example, that occupations which are attempting to become professions may be able to instill in their members strong pro fessional attitudes, while the more established professions may contain les idealistic members.

Variations from the professional model occur in two ways. In the firs place, occupations vary in the degree to which they are professionalized The established professions, such as medicine or law, appear to fit th professional model in most ways, although the attitudinal attributes ma or may not adhere to this pattern. The newer emerging professions d not appear to be as professionalized on the various attributes. If eac attribute is treated as a separate continuum, multiple variations are possi ble in terms of the degree to which occupations are professionalized.

The second form of variation is intraoccupational. Even among th established professions, members vary in their conformity to the profes sional model in both the structural and attitudinal attributes.[10] Both inter occupational and intra-occupational variations appear to be based o three factors. First is the general social structure which, at the abstrac level, may or may not need the services performed by the occupation and at the more pragmatic level, may not give the occupation the legal an behavioral sanctions to perform its functions. The second factor lie within the organization of the occupation itself. Here, for example, th presence of multiple and competing professional organizations may b divisive and thus inhibit professionalization through multiple standard for entrance and through varied regulative norms. The third source o variation is the setting in which the occupation is performed. The wor situation may have an impact on the degree to which the profession ca be self regulative and autonomous.

Professionals work in four distinct types of settings. Many members o established professions are found in individual private practice. This set ting for professional work appears to be diminishing in importance a organizationally based professional practice increases, and it will not b considered in this analysis. Lawyers are increasingly found in law firms o in legal departments of larger departments of larger organizations, an medical doctors are increasingly working in group or clinic practice Among the professionalizing groups, almost all are coming from an organ zational base such as the social work agency or the business firm.[11]

These organizational bases for professional occupations are of thre types. Following Scott's useful distinction, the first type is the *autonomou* professional organization exemplified by the medical clinic or the la firm.[12] Here the work of the professional is subject to his own, rathe than to external or administrative jurisdiction. The professionals then selves are the major determiners of the organizational structure, since they are the dominant source of authority. The second type is the *heteronomous* professional organization in which the professional employees are subordinated to an externally derived system. Examples are public schools, libraries, and social work agencies, all of which are affected by externally (often legislatively) based structuring. Scott suggests that the level of professional autonomy is correspondingly lessened in such a setting, a point which will be examined in this research. The third organizational setting is the professional *department* which is part of a larger organization. Examples of this are the legal or research departments of many organizations. In this kind of situation, the professionals employed are part of a larger organization and may or may not be able to affect the manner in which their own work is structured. It is this setting which has served as the basis for many discussions of professional-organizational conflict.

One way of analyzing these three types of settings is to determine the nature of the organizational structures found in the different organizational bases. In order to do this, the degree of bureaucratization within each type will be examined with a dimensional approach to the concept of bureaucracy.[13] This allows determination of the degree of bureaucratization of an organization, or a segment thereof, in terms of the degree of bureaucratization on *each* dimension.

The dimensions utilized are:

1. The hierarchy of authority—the extent to which the locus of decision making is prestructured by the organization.
2. Division of labor—the extent to which work tasks are subdivided by functional specialization decided by the organization.
3. Presence of rules—the degree to which the behavior of organizational members is subject to organizational control.
4. Procedural specifications—the extent to which organizational members must follow organizationally defined techniques in dealing with situations which they encounter.
5. Impersonality—the extent to which both organizational members and outsiders are treated without regard to individual qualities.
6. Technical competence—the extent to which organizationally defined "universalistic" standards are utilized in the personal selection and advancement process.

Each of these dimensions is treated as a separate continuum. Previous research has indicated that these continua do not necessarily vary together.[14] It is generally assumed that there is an inverse relationshi between professionalization and bureaucratization.[15] In this study th relationship will be examined without making that assumption. Instead, i is anticipated that the empirical findings will show that on some dimensions of bureaucracy, there is a positive relationship with professionalization. For example, a highly developed division of labor might well b related to a high degree of professionalization on all attributes, sinc professionals are specialists. By the same token, a high emphasis on technical competence as the basis for hiring and advancement would als appear to have a logical relationship with professionalization. On th other hand, a rigid hierarchy of authority seems incompatible with a hig level of professionalism, especially in terms of the attributes of autonom and colleague control. The presence of extensive organizationally base rules and procedures likewise appears to be negatively associated with high level of professionalization. Thus, findings of a mixed nature ar anticipated on this basis.

Methodology

Measurement of the degree of professionalization is accomplished in tw basic ways. The structural attributes of the several occupations include in the study will be examined using the method followed by Wilensky. Personnel managers and stockbrokers, occupations not included in h study, are examined here using the Wilensky format. The use of th approach not only allows comparison with the Wilensky work, but also w permit categorization of the occupations by length of time the profession attributes have been met by the occupation and the order in which the occupation has proceeded in its attempt to professionalize. According Wilensky, the closer the occupation follows a given sequence of professionalization steps, the more likely it is to be more professionalized. In th case, the two additional occupations would fall in the "doubtful professional" category, since some of the structural criteria are not met.[17]

The attitudinal attributes, including use of the professional organizati as a major reference, a belief in service to the public, belief in self regu tion, a sense of calling to the field, and a feeling of autonomy in work a measured by standard attitude scales. The scales were developed by u of the Likert technique.[18] This technique was chosen not only because its relevance for the kinds of attitudes being measured, but also because the question format is the same as that for the measurement of the degree of bureaucratization.[19]

Each of the bureaucratic dimensions was measured by means of a series of items which form a scale for the dimension.[20] The subjects in each organization responded to the items according to the degree to which the statement corresponded to their own perception of the organization. An ordinal scaling of each dimension was thus obtained. For both the professional attitude scales and the bureaucracy scales, the mean score for each set of respondents was utilized as the measure for the group involved.

At the outset of the research, it was decided to try to obtain a wide selection of occupations and employing organizations. The purpose of the research called for inclusion of occupations which are acknowledged professions in addition to those which are aspiring to become professions. At the same time, a variety of organizational settings was desired for the analysis of this variable. With these considerations in mind, organizations were contacted to determine if they were interested in cooperating in the research. The selection of organizations was based on at least some knowledge of their size and functioning. Those selected do not represent a sample of the universe of organizations, but they do appear to be representative of organizations of similar types. After the initial contact was made, personal interviews with officials in each organization were held and cooperation was solicited. Three organizations declined participation. Table 7-1 indicates the kinds of organizations included in the study, as well as the number of questionnaires distributed and returned. In all cases, all of the professional personnel were asked to complete the questionnaire.

The lawyers in the study represent three medium-sized law firms, two legal departments of private corporations, and part of a department of the federal government. The physicians are members of a medical department in a large government hospital and of a university student health service. The nurses work in a private general hospital and in the same university health service. Both of the Certified Public Accountant groups are part of large national CPA firms, while the accounting departments are part of larger organizations. The teachers are in the same school system at the levels indicated. The social workers represent two private agencies and a social work department of a school system. The stockbrokers are part of large national brokerage firms, while the advertising agencies are both small, regional concerns. The engineering department and the personnel departments are part of large national manufacturing firms. The library is the public library in a large metropolitan area.

Findings

Evidence from the behavior of the respondents provides strong support for the validity of two of the professionalism scales and also suggests that the attitudes measured are quite strongly associated with behavior. Back-

TABLE 7-1
Respondents' Distribution by Occupation, Type of Organization, Number Distributed, and Rate of Return

Occupation	*Type of organization*	*N distributed*	*N returned*	*%*
Physician	University Health	17	10	59
	Government Hospital	18	11	61
Nurse	University Student Health	15	8	53
	General Hospital	50	26	52
Accountant	University Department	3	2	67
	Manufacturing Firm	20	14	70
	CPA Firm 1	40	15	38
	CPA Firm 2	30	13	43
Teacher	High School	40	29	73
	Elementary School	14	12	86
Lawyer	Law Firm 1	18	9	50
	Law Firm 2	25	13	52
	Law Firm 3	13	9	69
	Legal Department 1	6	4	67
	Legal Department 2	17	11	65
	Legal Department 3	10	6	60
Social Worker	Private Agency 1	8	7	88
	Private Agency 2	17	11	65
	Public Agency	20	14	70
Stock Broker	Brokerage Firm 1	10	8	80
	Brokerage Firm 2	20	8	40
Librarian	Public Library	60	44	73
Engineer	Manufacturing Firm	20	15	75
Personnel Management	Manufacturing Firm 1	10	7	70
	Manufacturing Firm 2	25	14	56
Advertising	Advertising Firm 1	10	4	40
	Advertising Firm 2	6	4	67
Total		542	328	61

ground information regarding the frequency of attendance at professional meetings and membership in professional organizations was cross-tabulated with the attitude scale which measures the strength of the professional organization as a reference group. The behavioral data support the validity of the scales and also suggest that the respondents practice what they verbalize. Similarly, the scale measuring belief in self regulation is strongly associated with the actual presence of licenses or certification as reported in the background questions.[21]

The results on the attitude scales reveal some interesting and somewhat surprising patterns (see Table 7–2). On the belief in service to the public and sense of calling to the field attributes, both of which are related to a sense of dedication to the profession, the teachers, social workers, and nurses emerge as strongly professionalized. This may be related to the relatively low financial compensation which these fields receive since dedication seems necessary if one is to continue in the field. An interesting exception is that the teachers are somewhat weaker on the sense of calling to the field variable. An important factor appears to be the entry of many women into teaching because it is a "safe" women's occupation rather than because of any real dedictaion. The established professions are relatively weak on these variables.

Among the lawyers, the results indicate that on all but one attribute a legal department is more professionalized than the law firms. This suggests that professionals working in large organizations are not, by definition, confronted with situations which reduce the level of professionalization. The variations among the lawyers also suggest that the conditions of employment may play a dominant role in the development of professional attitudes. The lawyers in this study have quite common backgrounds in terms of their legal education and bar organizational memberships. Thus it appears that these attitudinal variations should probably be attributed to the organizational "climate" in which they work.[22]

Table 7–3 indicates the ranks for each occupational group on each of the bureaucratic dimensions. There are rather wide variations in degree of bureaucratization both among and within occupational groups. There are also rather strong interrelationships among the dimensions, as Table 7–4 indicates, except in the case of the technical competence dimension, where the relationship is reversed. Apparently these organizations, unlike more broadly based samples of organizations, are rather internally consistent in their degree of bureaucratization. It should be noted that these results represent both the total organization (in the cases of the autonomous and heteronomous professional organizations) and organizational segments (in

TABLE 7-2
Ranks on Professionalism Scales by Occupational Group

	Scale				
Occupational group	*(A) Professional organization as reference*	*(B) Belief in service to public*	*(C) Belief in self-regulation*	*(D) Sense of calling to field*	*(E) Feeling of autonomy*
Accounting					
CPA Firm 1	12	18	15	18.5	19
CPA Firm 2	16	10	20	10.5	6
Acct. Dept. 1	2.5	6	24.5	4	3
Acct. Dept. 2	1	1	2.5	1	1
Advertising					
Ad. Agency 1	8	4	9	15	20
Ad. Agency 2	20	23	7	20	23
Engineering					
Engineering Dept.	4	9	2.5	5	11
Lawyer					
Law Firm 1	22	16	19	10.5	14
Law Firm 2	13	14	22	7	22
Law Firm 3	23	12.5	18	13	15
Legal Dept. 1	26	11	23	8.5	10
Legal Dept. 2	99	24	14	2.5	25
Legal Dept. 3	11	5	4	2.5	26
Librarian					
Public Library	21	15	11	25	4.5
Nurse					
Nursing Division 1	10	20.5	16	18.5	4.5
Nursing Division 2	14	19	27	24	2
Personnel Mgmt.					
Per. Dept. 1	7	2	5.5	8.5	7.5
Per. Dept. 2	5.5	7.5	9	22	13
Physician					
Med. Dept. 1	19	3	17	21	21
Med. Dept. 2	5.5	7.5	9	16	12
Social Worker					
Private Agency 1	17	17	12	17	9
Private Agency 2	18	22	21	27	16
Public Agency	25	20.5	5.5	26	17
Stock Broker					
Firm 1	2.5	12.5	24.5	6	24
Firm 2	27	25	26	23	27
Teacher					
Elem. School	24	26	13	14	7.5
High School	15	27	1	12	18

Note: Lower rank indicates lower degree of professionalization.

TABLE 7-3
Ranks on Bureaucracy Scales by Occupational Group

Occupational group	*I* *Hierarchy of authority*	*II* *Division of labor*	*III* *Rules*	*IV* *Procedures*	*V* *Imper-sonality*	*VI* *Technical com-petence*
Accounting						
CPA Firm 1	3	5	5	6	9	16
CPA Firm 2	11	4	7	9	15.5	25
Acct. Dept. 1	19	25	21	26	24	2
Acct. Dept. 2	27	27	27	27	27	1
Advertising						
Ad. Agency 1	9.5	3	9	12	1	20.5
Ad. Agency 2	2	10.5	1	3	3	4
Engineering						
Engineering Dept.	17	18	17	19.5	11	3
Lawyer						
Law Firm 1	1	2	2	1	7	24
Law Firm 2	4.5	8	3	4	12	18
Law Firm 3	6	6	6	2	13	20.5
Legal Dept. 1	21	15.5	8	7	4	13.5
Legal Dept. 2	4.5	1	4	8	10	19
Legal Dept. 3	9.5	17	11	10	19.5	5
Librarian						
Public Library	25	19.5	26	24	14	23
Nurse						
Nursing Division 1	24	21	25	23	21.5	10
Nursing Division 2	26	23	24	25	25	15
Personnel Mgmt.						
Per. Dept. 1	18	26	15.5	19.5	26	8.5
Per. Dept. 2	22	13	18	18	6	12
Physician						
Med. Dept. 1	12	12	12	5	5	7
Med. Dept. 2	23	22	21	21	17	17
Social Worker						
Private Agency 1	16	10.5	13	13	21.5	22
Private Agency 2	14	15.5	14	16	8	26
Public Agency	13	9	10	14	2	13.5
Stock Broker						
Firm 1	7	14	22	22	18	6
Firm 2	8	7	15.5	11	15.5	27
Teacher						
Elem. School	20	19.5	23	17	19.5	11
High School	15	24	20	15	23	8.5

Note: Lower rank indicates lower degree of bureaucratization.

TABLE 7-4
Spearman Rank Order Correlation Coefficients between Bureaucratic Dimensions

	Hierarchy of authority	*Division of labor*	*Rules*	*Procedures*	*Imper-sonality*	*Technical competence*
Hierarchy of Authority	–	.791	.821	.737	.477	–.242
Division of Labor	–	–	.796	.782	.682	–.573
Rules	–	–	–	.898	.641	–.272
Procedures	–	–	–	–	.618	–.342
Impersonality	–	–	–	–	–	–.270

the cases of professional departments). Whether or not such internal consistency exists throughout the larger organizations involved in the latter case is subject to further investigation. It seems doubtful in view of the wider variations in task which would be found in the organizations as a whole. For example, while an engineering department might exhibit essentially the same degree of bureaucratization on the various dimensions, the sales department of the same organization might not.

The distribution of the groups into the autonomous, heteronomous, and departmental categories was relatively simple, using Scott's criteria. In the autonomous category are found the physicians (who, even though they are part of larger organizations, autonomously determine their own work structures), the law firms, the CPA firms, and the advertising agencies. The latter were included in this category because of their direct relationships with clients and the fee basis of their financing. Also, both agencies had been established by persons who had been in the advertising business with larger firms and had later established their own organizations without external administrative constraints.[23]

Similarly, the heteronomous organizations were rather easy to classify using Scott's criteria. The social work agencies, the library, and the two schools were placed in this category. Two other occupational groups also were placed in this category on the basis of their degree of self determination of structure and policy. Neither are considered by Scott in his development of this category, but both appear to fit. First are the nurses. As a professional group, nurses are subject to both the administrative and policy practices of the medical staff. Nursing services must, therefore adjudicate between the policies of the medical staff and their own professional codes. Stockbrokers also were placed in this category. This occupa ional grouping, which appears to operate quite autonomously within each egional office, nevertheless is subject to rather extensive external policies. The rules of the various stock and commodity exchanges, the Securities and Exchange Commission and the particular company policies themelves, appear to make placement in the heteronomous category most appropriate. Brokers do have individual clients and operate on a fee basis as do members of the autonomous category, but placement in this category appears unwarranted because of the factors noted above. The departmental category is comprised of the three legal departments, the engineering department, the personnel departments, and the accounting departments.

The data were analyzed by means of the Kruskal-Wallis one-way analyis of variance (H).[24] This technique allows the determination of whether or not the differences in the ranks of the various categories are due to chance or to real differences in the populations studied.

On the hierarchy of authority dimension, the autonomous organizations are significantly less bureaucratic than the other two types of occupational groupings, as Table 7–5 indicates. There is relatively little variation among the ranks in this category, while more variation exists within the heteronomous and departmental categories. This is consistent with Scott's suggesions in this regard. At the same time, the variations within the latter two categories are sufficient to suggest that neither category is inherently more rigid in its hierarchy of authority. With the exception of one legal department (see Table 7–3), the various settings in which lawyers work are essentially the same on this dimension. This suggests that work in the larger organization does not, by definition, impose a more rigid hierarchy on the practitioner. Nevertheless, the term "autonomous professional orga-

TABLE 7-5
Average Ranks and One Way Analysis of Variance (H) on Degree of Bureaucratization in These Types of Professional Settings

	Autonomy (average rank)	*Heteronomous (average rank)*	*Department (average rank)*	*H value (2 df)*
Hierarchy of Authority	8.0	16.8	17.4	6.96*
Division of Labor	8.1	16.3	17.9	7.31*
Rules	7.3	19.2	13.7	14.70***
Procedures	7.0	18.0	17.0	9.80**
Impersonality	9.2	16.8	15.9	4.29
Technical Competence	16.9	16.2	8.0	5.11*

* = $p < .05$.
** = $p < .01$.
*** = $p < .001$.

nization" appears to be relevant since these organizations do, in fac
exhibit less bureaucratization on this dimension.

On the division of labor dimension, the autonomous organizations ar
again much less bureaucratic, while the heteronomous organizations an
departments are essentially similar. The autonomous organizations exam
ined apparently have not begun the more intensive division of labor tha
Smigel found among larger New York law firms. On this dimension, als
the law firms and the legal departments do not vary widely among them
selves. While the division of labor is more intense in the other two cat
gories, again there is rather extensive variation in these categorie
Obviously, the extent of the division of labor is dependent upon the task
being performed rather than on the level of professionalization or th
externally imposed administrative structure.

Findings on the presence of rules dimension are somewhat consisten
with those already discussed. The autonomous organizations are chara
terized by much less bureaucratization on this dimension. In this cas
however, the professional departments are less bureaucratic than th
heteronomous organizations. This is consistent with Scott's discussio
but at the same time suggests that professional departments in larg
organizations are not inherently more bureaucratic than profession
organizations. The case of the lawyers in the two settings provides ev
dence for this position. These results raise an important issue in regard t
professional-organizational relationships. The data suggest that the occu
pational base of an organization (or an organizational segment) may hav
a real impact on the structure which the organization itself takes.[25] I
this case, the professional's self regulatory patterns may reduce the nee
for and utility of organizational rule systems. It might be hypothesize
that the more developed the normative system of the occupations in a
organization, the less need for a highly bureaucraticized organization
system.

Findings on the procedural specifications dimension of bureaucracy a
quite similar to those discussed above. The autonomous organizations a
characterized by quite a low level of bureaucratization, while the hete
onomous organizations are the most bureaucratic and the professional d
partments are slightly less bureaucratic.

On the impersonality dimension, the same general pattern emerge
although not to the same degree. Autonomous organizations are the lea
bureaucratic, while the other two types exhibit more impersonality. Th
higher level of impersonality found in the latter two categories may b

an aspect of the "bureaucratic personality" which has been discussed as a characteristic of organizations such as social work agencies. The aloofness and detachment which may inhibit the effectiveness of these groups appears to be due partially to the fact that the professions in the heteronomous category must deal with relatively large client populations. Those in the autonomous category have smaller such populations, and professions in the departmental category usually do not have individual clients. While the number of clients may contribute to impersonality, a high level of impersonality may itself inhibit the occupation in its drive toward professionalization. If public acceptance as a profession is a component of professionalism, then this higher level of impersonality may partially block further professionalization. The question remains whether or not this higher level of impersonality is a result of organizationally generated norms or of standards imposed upon the organization by the professions themselves.

The findings in regard to the last dimension, technical competence as the basis for hiring and advancement, are reversed from those previously discussed. On this dimension, the autonomous and heteronomous organizations are relatively bureaucratic, while the professional departments are relatively non-bureaucratic. Here it appears that a higher level of bureaucratization or more emphasis on technical competence is quite compatible with professional standards in that the practitioner is selected for employment and advancement on the basis of ability. In this case, a real source of conflict for persons employed in a professional department is evident if criteria other than performance are utilized in the personnel policies.[26]

The relationships between professionalization and bureaucratization are examined from two perspectives. First is an analysis of the relationships between the attitudinal variables and the bureaucratic dimensions, as indicated in Table 7–6. The generally negative relationships indicate that higher levels of professionalization are related to lower levels of bureaucratization, and vice versa.

When each of the professional variables is examined, some interesting patterns emerge. First, in the case of the professional organization as a reference group, there is a relatively small negative relationship between this variable and the presence of a rigid hierarchy of authority. It appears to make little difference if there is extensive reliance upon such a hierarchy in professionalized organizations. This conclusion is supported by the findings of Blau, et al., who suggest that the presence of such a hierarchy may facilitate the work of professionals if they serve coordination and

TABLE 7-6
Rank Order Correlation Coefficients between Professionalism Scales and Bureaucracy Scales

	Professional organization reference	Belief in service to public	Belief in self-regulation	Sense of calling to field	Feeling of autonomy
Hierarchy of Authority	−.029	−.262	−.149	.148	−.767**
Division of Labor	−.236	−.260	−.234	−.115	−.575**
Rules	−.144	−.121	−.107	.113	−.554
Procedures	−.360**	−.212	−.096	.000	−.603**
Impersonality	−.256	−.099	−.018	−.343*	−.489**
Technical Competence	.593**	.332*	.420**	.440**	.121

* = $p < .05$.
** = $p < .01$.

communication functions.[27] This is particularly so when the hierarchy is recognized as legitimate. The professional may thus recognize and essentially approve of the fact that certain decisions must be made by people in the hierarchy.

A stronger negative relationship is found on the division of labor dimension. If a division of labor is very intense, it may force a professional person away from his broader professional ties. This interpretation recognizes specialization within the professions, but the question here is the level of organizationally based division of labor. At the same time, strong professional identification may impede intensive specialization on the part of organizations. A weaker relationship is found on the presence of rules dimension. Organizationally developed rules governing the behavior of members appear not to intrude strongly on this or on other professional attitudes.

There is a strong negative relationship between professional attitude and the procedural specifications dimension. This is predictable since strong professional orientations appear to be in basic conflict with organizationally developed techniques of dealing with work situations. As more procedures are developed by the organization, they may become more burdensome for the professional.

The previously discussed relationship between the level of professionalization and the degree to which impersonality is stressed is borne out by the findings here in regard to the impersonality dimension. That is the more professional the attitude on this variable, the less impersonality is stressed. The more professional groups apparently do not need to utilize impersonality in their organizational arrangements.

The strong positive relationship between this professional variable and the organizational emphasis on technical competence is not unexpected. Since this bureaucratic dimension is so strongly related to most of the professional attitudes, it might serve as an informal indicator of the level of professionalism in organizations if other indicators are not available.

The findings on the belief in service to the public, belief in self-regulation, and sense of calling to the field variables are essentially similar to those just discussed. The areas of congruence and conflict which might emerge also are similar.

Strong negative relationships exist between the autonomy variable and the first five bureaucratic dimensions. This suggests that increased bureaucratization threatens professional autonomy. It is in these relationships that a potential source of conflict between the professional and the organization can be found. The strong drive for autonomy on the part of a professional may come into direct conflict with organizationally based job requirements. At the same time, the organization may be threatened by strong professional desires on the part of at least some of its members. Future research should delineate both the extent of the conflict and its sources and the extent to which it is felt by and threatens both the professional(s) and the organization.

When the structural aspect of professionalization is considered, essentially the same findings emerge. As Table 7-7 suggests, the more professionalized groups are found in settings which are less bureaucratic. The more professionalized groups, that is, those with more self regulation and longer socialization in preparation for the field, perhaps do not "need" the same kinds of organizational controls as less professionalized groups in

TABLE 7-7
Average Ranks and One Way Analysis of Variance (H) on Degree of Bureaucratization of the Work Settings of Three Types of Professionals

	Established (average rank)	*Process (average rank)*	*Doubtful (average rank)*	*H value (2 df)*
Hierarchy of Authority	9.5	18.9	14.1	6.14*
Division of Labor	9.2	17.8	15.7	5.57
Rules	7.9	19.1	15.9	9.65**
Procedures	7.3	18.5	17.3	10.96**
Impersonality	11.2	16.2	15.1	1.65
Technical Competence	16.5	14.7	10.2	2.02

* = $p < .05$.
** = $p < .001$.

dealing with problems and decisions. At the same time, the presence o more bureaucratic systems for the less professionalized groups may serv as an inhibitor to their further professionalization.

Summary and Conclusions

Among the major findings of this research is the fact that the structura and the attitudinal aspects of professionalization do not necessarily var together. Some "established" professions have rather weakly develope professional attitudes, while some of the less professionalized groups hav very strong attitudes in this regard. The strength of these attitudes appear to be based on the kind of socialization which has taken place both in th profession's training program and in the work itself. An additional facto is the place of the occupation in the wider social structure. If the occupa tion receives relatively few rewards in a material sense, the level of dedica tion is likely to be higher. If the occupation is allowed to be self-regulat ing, it will tend to believe quite strongly in this. Therefore, changes i the social structure may bring about corresponding attitudinal adjust ments.

The organizations in which professionals work vary rather widely i their degree of bureaucratization. The variation is not based on the dis tinction between professional departments and professional organization since some professional departments are less bureaucratic than some pr fessional organizations, and vice versa. There is, however, a general ten ency for the autonomous professional organization to be less bureaucrat than either the heteronomous organization or the professional departmen This suggests that the nature of the occupational groups in an organizatio affects the organizational structure. The workers (professionals) *impor* standards into the organization to which the organization must adjust. I the development of a new organization, this importation would probabl occur without any conflict. In an established organization, the importa tion, either by an entire department or by new employees within a profe sional department, might be a real source of conflict if the professiona and organizational standards do not coincide.

With the exception of the technical competence dimension, a generall inverse relationship exists between the levels of bureaucratization an professionalization. Autonomy, as an important professional attribute, most strongly inversely related to bureaucratization. The other variable are not as inversely related. This suggests that increased bureaucratization and professionalization might lead to conflict in either the professional organization or department, but that this conflict is not inherent, given the relative weakness of most of the relationships found. Conflict occurs within a professional group or within an organization only to the degree that specific aspects of bureaucratization or professionalization vary enough to conflict with other specific aspects. Stated in another way, the implication is that in some cases an equilibrium may exist between the levels of professionalization and bureaucratization in the sense that a particular level of professionalization may require a certain level of bureaucratization to maintain social control. Too little bureaucratization may lead to too many undefined operational areas if the profession itself has not developed operational standards for these areas. By the same token, conflict may ensue if the equilibrium is upset.

An assumption of inherent conflict between the professional or the professional group and the employing organization appears to be unwarranted. If it is present, the bases of conflict in terms of professional attitudes and/or organizational structure should be made explicit. After this is done, the analysis of conflicts based upon specific issues, such as resistance to non-professional supervision, can proceed. After any particular conflict situation, changes in either the professional orientations or the organizational structure should be noted so that the essentially stable conditions from which conflicts emerge can be established. Since conflicts probably create changes, any conflicts which follow would necessarily emerge from a different setting than that originally noted. These changes must be noted in any longitudinal analyses of professional-organizational relations.

NOTES

1. See Howard M. Vollmer and Donald L. Mills (eds.) *Professionalization*, Englewood Cliffs, N.J., Prentice-Hall, 1966, for a clear distinction between the terms profession, professionalization, and professionalism (pp. vii–viii). Their distinctions will be followed in this paper.

2. Harold L. Wilensky, "The Professionalization of Everyone?" *American Journal of Sociology*, Vol. LXX, September, 1964, pp. 137–158.

3. Theodore Caplow uses a slightly altered formulation in his "Sequential Steps in Professionalization," in Vollmer and Mills, *op. cit.* pp. 20–21. Both Wilensky and Caplow include the same general variables, with Wilensky's appearing to be more descriptively accurate.

4. For discussions of the importance of autonomy see William Kornhauser, *Scientists in Industry*, Berkeley, University of California Press, 1963 and Simon Marcson, *The Scientist in American Industry*, New York: Harper Bros., 1960.

5. See William J. Goode, "Community Within a Community: The Professions," *American Sociological Review*, Vol. 22, April, 1957, p. 194; and Greenwood, *op. cit.*

6. See Edward A. Gross, *Work and Society*, New York: Thomas Y. Crowell Co., 1958, pp. 77–82; and Parsons, *loc. cit.*

7. See Greenwood, *op. cit.*, and Goode, *op. cit.*

8. See Gross, *op. cit.*

9. Richard Scott distinguishes between autonomous and heteronomous professional organizations while, at the same time, stressing the importance of autonomy for the professional in "Reactions to Supervision in a Heteronomous Professional Organization," *Administrative Science Quarterly*, Vol. 10, June, 1965, pp. 65–81.

10. Bucher and Strauss, in "Professions in Process," *American Journal of Sociology*, Vol. LXVI, January, 1961, pp. 325–334 discuss this type of variation as do Jack Ladinsky and Joel B. Grossman in "Organizational Consequences of Professional Consensus: Lawyers and Selection of Judges," *Administrative Science Quarterly*, Vol. 11, June, 1966, pp. 79–106.

11. Brandeis recognized the trend toward professionalization among businessmen as early as 1925 in Louis D. Brandeis, *Business—A Profession*, Boston: Small, Maynard, 1925.

12. W. Richard Scott, *op. cit.*, pp. 65–81.

13. See Stanley H. Udy, Jr., "Bureaucracy and Rationality in Weber's Theory," *American Sociological Review*, Vol. 24, December, 1959, pp. 791–795, and Richard H. Hall, "The Concept of Bureaucracy: An Empirical Assessment," *American Journal of Sociology*, Vol. LXIX, July, 1963, pp. 32–40. The discussions are primarily based, of course, on Weber's original discussion of bureaucracy in Max Weber, *The Theory of Economic and Social Organization* (trans. A. M. Henderson and Talcott Parsons), New York: Oxford University Press, 1947.

14. Hall, *op. cit.*

15. Peter M. Blau, Wolf V. Heydebrand, and Robert E. Stauffer in "The Structure of Small Bureaucracies," *American Sociological Review*, Vol. 31, April, 1966, pp. 179–191 began with this assumption which was later modified.

16. Wilensky, *op. cit.*

17. William M. Evan discusses some of the barriers to the professionalization of stockbrokers in "Status-Set and Role-Set Conflicts of the Stockbroker: A Problem in the Sociology of Law," *Social Forces*, Vol. 45, September, 1966, pp. 80–82.

18. The advantages of this form of scaling are discussed in John E. Barclay and Herbert B. Weaver, "Comparative Reliabilities and Ease of Construction of Thurstone and Likert Attitude Scales," *Journal of Social Psychology*, Vol. 58, 1962, pp. 109–120 and Charles R. Tittle, "Attitude Measurement and Prediction of Behavior: An Evaluation of Five Measuring Techniques," unpublished Ph.D. Thesis, University of Texas 1965.

19. Each scale attained a reliability of .80 or higher using the split-half method with the Spearman-Brown correction formula. Pretest data also suggested that the scales were valid. Physicians, nurses, teachers, and accountants comprised the pretest group A complete discussion of the procedures followed is contained in Richard H. Hall, *Professionalization and Bureaucratization*, unpublished manuscript, Indiana University, 1966

20. Hall, "Concept of Bureaucracy," *loc. cit.*, contains a discussion of the development of these scales.

21. Those who strongly believed in using the professional organization as a reference group also belonged to and attended the meetings of professional organizations ($p. < .001$). Similarly, belief in self regulation is associated with the presence of licensing and certification ($p. < .001$).

22. Jerome E. Carlin suggests that law firms vary in the degree to which ethical norm

are enforced. Jerome E. Carlin, *Lawyers' Ethics*, New York: Russell Sage Foundation, 1966, pp. 96–103. The climate should also affect other professional attitudes.

23. See Scott, *op. cit.*

24. For a discussion of this technique, see Sidney Siegel, *Non-Parametric Statistics for the Behavioral Sciences*, New York: McGraw-Hill, 1956, pp. 184–193.

25. Blau, *et al.*, *op. cit.*, and D. S. Pugh, D. J. Hickson, *et al.*, "A Scheme for Organizational Analysis," *Administrative Science Quarterly*, Vol. 8, December, 1963, pp. 289–315 make this suggestion, also.

26. Marcson has noted the frustrations felt by scientists when the reward system is blocked if they follow a strictly research route in their careers. The higher rewards come from joining the administrative structure, which is a contradiction of their professional standards. Simon Marcson, *The Scientist in American Industry*, New York: Harper Bros., 1960, pp. 66–71.

27. Blau, *et al.*, *op. cit.*

8

GEORGE BRAGER

COMMITMENT AND CONFLICT IN A NORMATIVE ORGANIZATION

Conflict theorists posit that an external threat to a collectivity results in a tightening of the internal bonds of solidarity. The classic position has been stated by Sumner (1906:12–13): "The relation of comradeship and peace in the we-group and that of hostility and war towards others-group are correlative of each other. The exigencies of war with outsiders are what make peace inside."

Coser (1961:87) similarly notes that "conflict with out-groups increases internal cohesion." He suggests that "outside conflict . . . mobilizes the group's defenses, among which is the reaffirmation of their value system against the outside agency" (1961:90).

We suspect however, that in-group cohesion is not so frequent a response to external threat as the literature has suggested. It may be, in fact, that external threat leads to dissensus or internal conflict just as often, and that the structure of some collectivities predisposes this outcome. Studies of the effects of external conflict upon collectivities tend to ignore differences among these bodies. Empirical evidence of responses to conflict derives largely from small group research, while theorists have been interested in large-scale systems, such as nation-states. The effects of external conflict upon complex organizations have been largely ignored.

This paper reports upon an empirical study of an organization which was subjected to virulent outside attack and consequently rent by internal conflict. The research attempts to relate the disharmony which developed regarding the organization's strategy to counteract attack to the structurally based value commitments of the participants. The model of a norma-

Reprinted from George Brager, "Commitment and Conflict in a Normative Organization," *American Sociological Review*, 34 (August, 1969), pp. 482–491.

tive organization, drawn from the work of Etzioni (1961, 1964, 1965), serves as the study's basis for analysis.

The Setting: Mobilization for Youth, a Normative Organization

Mobilization for Youth (MFY), a comprehensive delinquency-prevention and anti-poverty project on New York's Lower East Side, is the setting for the study.[1]

On June 1, 1962, the front page of the *New York Times* carried President Kennedy's announcement that the project had received grants of $12.6 million from federal, city and foundation sources. Less than 27 months later, Mobilization was once again on the front pages of the New York press, this time the subject of very nearly fatal attack. On August 16, 1964, during debate in Congress regarding the passage of the Economic Opportunities Act, the *New York Daily News* ran the headline: "Youth Agency Eyed for Reds—City Cuts Off Project Funds." The article accused the demonstration project of using "its facilities—and juveniles—to foment rent strikes and racial disorders." (The attack occurred one month after riots in Harlem and Bedford-Stuyvesant). During the following five months, MFY was subjected to repeated attacks in the press and to investigation by city, state, and federal, as well as internal, sources. It acted and reacted under the buffeting of many of the most powerful groups in the city.

In structure, MFY is a non-profit membership corporation which follows the usual pattern of a philanthropic service agency. At the time of the attack, it had a governing board of directors representing a variety of interest groups and backgrounds.[2] Its staff was organized in hierarchical fashion and included three major subgroupings: (1) an executive staff which consisted of the top leadership of the agency's major program divisions, (2) supervisory and consultative personnel of varied professional backgrounds, and (3) practitioners, including both professionals and non-professionals, who worked in a program capacity directly with agency clients. There was, in addition, a practitioner group of "caretakers," who fulfilled such instrumental functions as public relations, finance, and administration.

MFY is a normative organization in the Etzioni schema. Normative organizations are characterized by the use of identitive power (i.e., the power to make people identify with the organization) as its major source of control and by high commitment on the part of participants. Etzioni argues that these organizations which serve culture-oriented goals (e.g., religious institutions, ideological-political groups, colleges and universities, and voluntary associations), have to rely on identative powers since the realization of their goals requires positive and intense commitment on the part of participants, and such commitments cannot effectively be attained by other means.

Although the Etzioni schema would lead us to predict widespread commitment by participants to the values of normative organizations such as MFY, the classification system is too crude for precise analysis of a single organization. Etzioni posits that the basis of involvement—whether social, normative, or utilitarian—affects both a participant's commitment and an organization's reward system. But the positive commitment of participants in normative organizations may stem from a social base, a normative one, or both. Further, the consequences of utilitarian rewards in normative organizations are unclear, e.g., whether the application of utilitarian incentives makes appeals to "idealistic" (normative) motives less effective or whether it intensifies value commitment. If, as may logically be argued, the basis of involvement also varies with rank and function, there will be differential commitments within normative organizations, and these will be structurally determined.

It is often assumed, for example, that persons who serve on the boards of directors of welfare agencies do so for normative reasons. Opinion studies, however, indicate sharp disparity between board members' positions on various issues and those of professionals, with lay officials in some instances even challenging the need for agency services (Kramer, 1965; Boehm, 1964). Board members of social agencies have a tendency toward social involvement, and there is even some evidence that utilitarian involvement may be more widespread among occupants of voluntary, nonremunerative positions than has been assumed (Ross, 1954). In any case, to include board members of such potentially varying commitments into a single classification scheme may obscure important differences.

Responding to External Threat

Organizations are responsive to varying publics, external and internal. One consequence of a public attack is that it upsets the equilibrium with which the organization has in the past accommodated to these varying publics.

Actions invisible in the past now become observable, and formerly uninterested groups now become concerned. (Invisibility permits leaders considerable latitude in choosing the members of the organizational coalition to whom they will be responsive.) More importantly, power balances shift as different groups assume greater or lesser salience as a point of reference for agency decision-making. In short, public attention constricts an organization's freedom of action and encourages greater responsiveness to now-attentive others.

There is evidence that MFY's staff perceived this greater responsiveness to its external publics during the organization's crisis. They believed, for example, that the organization did not sufficiently consider their opinion in deciding upon crisis strategies.[3] They saw the board as too responsive to city officials and unresponsive to staff needs.[4] They believed that important principles were compromised in the handling of the attack.[5] According to MFY's staff, as powerful external adversaries became the primary point in decision-making, the agency neglected its internal constituency.

Conflict involving normative organizations is likely to be extremely intense (Aubert, 1963). An organization facing a challenge to its value base and a threat to its survival may defend itself in two distinctly different ways: (1) to defend the organization's values; (2) to focus upon survival to the exclusion (if necessary) of values. Strategy may move fitfully from one of these considerations to the other and ultimately end somewhere between the two polar extremes. Nevertheless, the threat of organizational extinction will inevitably turn the focus sharply in the direction of organizational maintenance, and some compromise on values is often deemed a requirement in developing a defense against outside attack. So, for example, on the day following the attack, MFY's executive directors assured the staff that no one would be fired for suspected communist affiliation. "What matters," they said, "is not really whether we survive, and that indeed is an open question—what matters is the issues on which we stand and fight . . ." (Tape of meeting, 9/17/64). Three days later, however, in response to pressure from federal officials who believed that MFY could not otherwise be saved, the executive directors agreed that *current* members of "subversive" organizations would be dismissed.

The reactions of participants to an organizationally-protective defense posture will be determined by the intensity of their commitment to those organizational values which are threatened (i.e., by the degree to which they are normatively involved). Value committed participants are more likely to urge an activist defense of the organization's ideology. If they perceive organizational strategy as a compromise, they will react negatively. Those motivated predominantly by social and utilitarian involvement will be less committed to the defense of the organization's values and more receptive to the defense of the organization as an instrumentality. They are more likely, in other words, to protect its interests and compromise its beliefs.

We note an important qualification to the above. It may be that value-committed participants with a long time perspective will espouse a survival strategy and value compromise so that the organization may endure to fight another day. In that case, they could be as normatively committed as other participants, the difference being their long time perspective. Upper echelon members are more likely than lower ones to hold this long time perspective. Committed persons with a short time perspective, on the other hand, are more likely to view a survival strategy as a sacrifice of principle. As shall become clear subsequently, however, we do not believe that a differing time perspective accounts for the varying reactions of Mobilization's participants to the organization's strategy.

Whether the espousal of an organizationally-protective defense posture stems from a long time perspective or lesser value commitment, however, contention between participants who counsel this position and those who urge an activist defense is a logical consequence. Furthermore, if, participants perceive themselves as having sufficient influence to affect the course of events, virulent internal conflict is highly possible. In any case, dissensus may be the expected response to attacks (1) which endanger organizational values (and this is virtually inevitable with normative organizations), (2) in which organizational defense entails compromise (the ordinarily favored strategy), and (3) when commitment to the organization's values is high but is differentially distributed among participants (and particularly among those with access to influence).

In order to assess the sources of internal discord in MFY, we must note an important qualification in the literature to the position that external conflict breeds internal solidarity. This relates to the initial cohesiveness of the group as it faces outer conflict. Thus, Coser (1961:93) suggests that "the relation between outer conflict and inner cohesion does not hold true where internal cohesion before the outbreak of the attack is so low that group members have ceased to regard the preservation of the group as worthwhile, or actually see the outside threat to concern 'them' rather than 'us.'" Our first hypothesis, then, examines whether organizational values were widely diffused among MFY's participants. If so, the internal conflict which wracked the organization cannot be explained by the belief that outside threat concerned "them" rather than "us."

Our second hypothesis examines whether the degree of commitment to MFY varied with the rank and function of participants. If confirmed, a point essential to our exploration is made—namely, that differences in value commitment are associated with organizational structure, and are not solely the reflection of an actor's personality or character.

As already suggested, we predict that differences in the initial degree of value commitment are likely to have consequences for the development of dissensus. It may be, however, the stress engendered by the attack, rather than the commitments of the participants, that explains MFY's internal discord. The point is important, since it constitutes the second major qualification in the literature to the outer-conflict, inner-cohesion expectation. Thus, Janowitz (1959:73), on the basis of his study of the military, concludes that continued exposure to stress weakens primary-group solidarity and undermines organizational effectiveness.

We test, finally, for a relationship between an actor's dissatisfaction with MFY's crisis strategy and his role in decision-making. We do so because we suspect that the influence relations of actors in dissensus may importantly determine whether potential civil conflict becomes a reality. A perception of influence may be a precondition of actual conflict, since individual actors or constituent groups are likely to act only if they perceive the possibility of affecting the course of events.

The following, then, are our hypotheses:

1. *Commitment to MFY's values will be widely diffused among its participants.* Operationally, we predict a positive response to a Commitment to Values measure, which assesses opinions regarding MFY's controversial social change orientation.[6]

2. *Commitment to Mobilization's values will vary with the rank and function of the participant.* This hypothesis is operationalized by examining whether there are differential responses to the Commitment to Values Index by participants of differing hierarchical levels and organizational function.

3. *Internal dissensus resulting from the external attack upon MFY will be related to the participants' ideological commitments to the organization.* We predict a relationship between responses to the Commitment to Values measure and a Dissensus score. It is expected that the higher the commitment, the more negative the reaction to the organization's defense strategy.

4. *There will be a relationship between the amount of stress experienced by participants during the attack and their evaluation of crisis strategy.* We predict, specifically, an association between respondents' scores on the Dissensus measure and their response to items which assess personal stress. Those high in stress, we expect, will be more negative about the agency's posture in defense.

5. *There will be a relationship between participants' role in decision-making and their evaluation of crisis strategy.* Scores on the Dissensus measure will, we predict, vary with a person's cumulative score on a series of ten items, each representing an area of decision-making for which the respondent reports that he had or did not have responsibility.

Method

The research instrument was administered to the MFY board and staff six months following the "settlement" of the dispute, with a completion rate of 76% and 74%, respectively.[7]

To measure the extent of commitment to the value and goals of MFY, the main values of the organization had to be isolated and respondents asked to indicate the extent to which they subscribed to these values. MFY's value orientation is officially set forth in two major documents: the proposal to its funding sources (Mobilization for Youth, 1961), and an extensive interim report of its action programs (Mobilization for Youth, 1964), submitted to the same parties after two years of operation. The two documents are consistent with respect to MFY's values, with the two-year report being particularly relevant as a source for the Commitment to Values Index since it was completed barely two months before the attack upon the organization. It was written by staff at varying hierarchical levels, reviewed by the heads of relevant program divisions, adopted by the board, and submitted to the funding sources. It serves exceedingly well as a statement of the operative value position of the organization at the time of the crisis.

The items which comprise the Index are for the most part direct quotes from the proposal and the interim report, modified to fit an agree-disagree questionnaire format. Respondents were asked their position on a five-point continuum from "strongly agree" to "strongly disagree" (Likert scale) for the 20 items of the index. Their ranks on these items were then summed to form a composite score for each respondent. A high score implies more commitment and a low score less commitment to MFY's values and goals.

Although the items were deemed to have face validity, the names of high and low scorers were submitted to a panel of judges acquainted with the organization's personnel. A significant level of discrimination for the composite index score was obtained at the .01 level. The reliability of the scale, as measured by the split-half method with the Sperman-Brown correction for attenuation, was .771.

The research instrument also includes 15 items which, taken together, indicate the respondents' evaluation of MFY's defense during the crisis. We have used this as our measure of dissensus, since it reflects the participants' disagreement with the official positions of the organization.

During the months of the organization's travail, there were, broadly speaking, three major issues. One related to the extent to which the board had lost the initiative for MFY's affairs. Some participants objected to what was described as compliance strategy—the board's unwillingness to mount a frontal attack upon those who were deemed its attackers. A second issue stemmed from tensions in board-staff relations; e.g., was the board appropriately responsive to staff opinion? A third issue was the extent to which board and executive staff were willing, in their concern for the organization's survival, to compromise basic principles in regard to either organizational values or the civil liberties of staff accused of "subversion."

The Dissensus (D) Scale is composed of opinion statements reflecting these issues. Respondents were scored from 1 to 4 on each, depending upon their degree of agreement with the items. Eight of the 15 items were worded so that agreement meant support of Mobilization's crisis strategy. These were reversed in scoring. Thus, high scorers were negative perceivers of the crisis posture, while low scorers agreed with the agency's policies.

Since the D scale is a series of opinion statements, no external criteria exist against which to correlate the respondents' subjective statements. The items were, however, checked for consistency. To cite one example, an overall opinion question ("Mobilization's defense against the attack was as effective as could be expected under the circumstances") was included in the instrument. It was predicted, and confirmed, that the attitude of the respondent, as reflected in his D score, would be closely related to his overall opinion regarding defense effectiveness, as indicated by his response to the single item. (The correlation between the two was .725.) The reliability of the D score, as measured by the split-half method with correction, was .948.

In addition to commitment and reaction to crisis strategy, the instrument attempts to identify those participants who were involved in organizational decision-making. This was tested by asking participants whether or not they had been involved in making decisions in each of the ten operational areas (e.g., hiring and firing of staff, allocation of funds, program content, etc.). No attempt was made to assign different weights to these various areas of decision since competent raters might disagree as to their relative importance. The actor's "influence score" was determined by assigning one point for each decision-making area and totalling the items checked.[8]

It was assumed that the predictive validity of the score could be tested by determining whether the results discriminated the hierarchical locations of Mobilization's constituent groupings. Logically, it was expected that, if the index was valid, the executive, supervisory, and practitioner staff would score in that order. This is in fact what occurred.

Finally, a combination of two questions was used to operationalize the notion of stress. Participants were asked how, at the time of the crisis, they had appraised the likelihood of their being attacked for their political activities. Possible responses were "very good," "good," "some," and "none." This was followed by the query, "Suppose you *had* thought there was a possibility that you would be publicly attacked, would you have been concerned?" Respondents were given four choices: unconcerned, mildly concerned, very concerned, and extremely concerned. Those who checked either of the last two alternatives *and* who thought there was a "good" or "very good" chance that they personally would be attacked for political activities were deemed to have been under stress during the attack.

We assumed that our measures of stress would be valid if they correlated with actual vulnerability to attack. It might reasonably be predicted that those who either were publicly accused of "subversion" or were in real danger of such accusation would feel considerable stress in such circumstance. The prediction is confirmed: 15 of the 21 respondents whose names came to MFY officials for suspected "subversive" activity (71.4%) fall into the stress category and six do not (28.6%), as compared to 49.3% of other staff who felt stressful and 50.7% who did not.

Findings

MFY's values, although controversial, appear to have been widely accepted by the organization's participants. Unfortunately, measures do not exist with which to compare the commitment of participants in other types of organization to the values of those organizations. Our findings regarding the diffusion of values in Mobilization for Youth are not comparative, and must therefore remain suggestive rather than definitive.

Theoretically, each respondent could have obtained a total score on the Commitment to Values Index ranging from 20 to 100. The actual scores obtained, however, are sharply skewed in the positive direction: 121 respondents (53.1%) scored 80 and above, and 59 respondents (25.9%) ranged from 70 to 79. Only 11 persons, or 4.8% of the total sample, scored between 50 and 59, and no one obtained a score lower than 50.[9]

Our second hypothesis is confirmed—the degree of commitment to MFY's values is differentially distributed by hierarchical level and organizational function. The percentage of executive staff in the upper third commitment-to-values group (48.1%) exceeds the corresponding percentage of supervisory and consultant staff (38.4%), while the percentage of supervisors in the upper third exceeds that of practitioners (26.1%). Of special interest is the rank in this ordering of the board of directors, the group which is in the superordinate, decision-making position. The percentage of board members scoring in the upper third (10.7%) is even lower than the corresponding percentage of practitioners. The result is unexpected when we consider that the statements which comprise the Commitment to Values Index were drawn from documents which required and obtained board approval.

Organizational function also appears to have affected commitment to agency values. Program staff, whose functions are goal (and normatively) related, manifest dramatically more commitment than staff whose functions are instrumental (managerial, financial, public relations). None of those who performed managerial functions were in the highest third commitment group, as compared to 30.9% of all program staff. Interestingly, MFY's board fell midway between the program and managerial staff on commitment to agency values. The latter finding suggests that the dual function of a board (normative and managerial) predisposes a dual outlook.

What, then, is the relationship between commitment to the organization's values and the evaluation of its strategies to counter outside attack? As shown in Table 8–1, our third hypothesis is confirmed.

There is a striking relationship between high commitment to agency values and negative evaluation of its strategies. Thus, 54.1% of MFY's most committed participants were negative reactors, as compared to 13.1% who reacted positively. Among the least committed Mobilization participants, the figures are reversed.

Our fourth hypothesis predicts that dissensus will be associated with stress. In MFY's instance, this was not the case; the relationship between a feeling of stress and evaluation of the agency's course was not statistically significant. In order to probe the data, the Commitment to Values Index and the D scale were dichotomized: the first into high and low commitment categories, the latter into positive and negative crisis reaction. Table 8–2 explores the effect of stress upon crisis reaction while controlling for commitment to Mobilization's values.

The differences between less committed participants who felt stress and those who did not is worth noting. Less committed and non-stressful participants tended toward more positive crisis reaction (77.1%) than less committed but more stressful participants (46.2%). On the other hand, there is little difference in crisis reaction between those with and without stress (77.5% and 72.7%, respectively) among the more highly committed participants. Thus stress appears to be a more salient factor in dissensus among the less committed than among those who are more so.

We turn now to our final hypothesis, the relationship between participants' role in decision-making and their evaluation of crisis strategy. Logically, it is to be expected that those who perceive themselves as

TABLE 8-1
Reaction to Crisis Strategy by Commitment to Values
(N = 207*; Board of Directors and All Program Staff)

Commitment to values index	Reaction to crisis strategy				
	Approval	*Mid-position*	*Disapproval*	*Total*	*Percent*
Low	57.4% (35)	37.7% (23)	4.9% (3)	— 61	— 29.5
Medium	27.1% (23)	29.4% (25)	43.5% (37)	— 85	— 41.0
High	13.1% (8)	32.8% (20)	54.1% (33)	— 61	— 29.5
Total	66	68	73	207	100.0

*Excludes managerial staff and 9 "no response."

TABLE 8-2
Reaction to Crisis Strategy by Individual Stress, Controlled for Commitment to Values (N = 134*, Program Staff Prior to Attack)

	Commitment to values index							
	Low commitment				*High commitment*			
	Reactions to crisis strategy							
Stress	*Approval*	*Disapproval*	*Total*	*Percent*	*Approval*	*Disapproval*	*Total*	*Percent*
No stress	77.1%	22.9%	—	—	27.3%	72.7%	—	—
	(27)	(8)	35	57.4	(9)	(24)	33	45.2
Stress	46.2	53.8	—	—	22.5	77.5	—	—
	(12)	(14)	26	42.6	(9)	(31)	40	54.8
Total			61	100.0			73	100.0

*Excludes one "no response."

organizational decision-makers will be most likely to approve an organization's course in fending off attack. Responsibility, we might suppose, would generate attachment to the organization as an entity. In part, also, negative judgments regarding the policies of an organization implies negative judgment about its decision makers.

In MFY's instance, however, the expected finding is reversed—the greater the participant's involvement in decision making, the more negative his reaction to official crisis strategy. How are we to account for this striking reversal? One possible explanation lies in the relationship between a negative reaction to crisis strategy and commitment to MFY's values, on the one hand, and commitment to MFY's values and decision-making involvement, on the other. To aid in examining this possibility, we have dichotomized the three variables: commitment into a high and low category, crisis reaction into positive and negative response, and decision-making involvement into none and some. When we controlled for commitment, the significance of the relationship between decision-involvement and crisis reaction disappeared. Among the most committed participants, decision-making approached, but did not reach, significance (at the .09 level). Among the least committed members, no difference in crisis reaction was evident on the part of those involved in none or some decision-making. We conclude, therefore, that the negative reaction to strategy of MFY decision-makers stemmed from their high commitment to the organization's values, rather than from decision-involvement. Table 8–3, however, shows an interesting relationship among the three variables.

It indicates that among negative crisis reactors, those who were less

TABLE 8-3*
Commitment to Values by Decision Involvement for Respondents with Negative Reaction to Crisis Strategy

	Negative reactors to crisis strategy		
	Commitment to values index		
Decision-making involvement	*Low commitment*	*High commitment*	*Difference*
None	32.4%	58.3%	25.9
Some	41.9	79.7	37.8
Difference	9.5	21.4	–

*This table is a composite of the negative reaction portion of the two contingency tables which were used to test decision-involvement controlled for commitment.

committed appeared to be less affected by their participation in decision-making than the more committed members. Thus, the percentage point difference within the low-commitment group, depending upon whether they had some decision-making responsibility or none, was only 9.5. This contrasts with their more committed colleagues, who showed a 21.4 difference as a result of some decision-involvement or none. Furthermore, among negative reactors who are highly committed, involvement in decision-making appears to increase the likelihood of their holding a critical view of the organization's defense.

Interpretation

Caution must be exercised in generalizing from the MFY experience. Study of a single organization can yield insights and information, but there is no assurance that the data are not unique to that organization. We can only assert that discord within MFY following an external attack was *not* caused by organizational malintegration and lack of cohesion, or by excessive strain upon the participants. Since these are major explanations found in the literature, their non-applicability in the MFY experience suggests the need for attention to other as yet unexplored causes.

It is clear that the pattern of conflict in MFY is related to the *particular* course pursued by the agency to defend itself. We have already argued, however, that Mobilization's defense stance was more usual than unique. Its responsiveness to powerful external publics and its focus upon the protection of the organization as an instrumentality, at least as perceived by MFY participants, is the expected response of an organization whose survival is threatened by outside attack.

In any event, the MFY data suggest that when values are perceived as compromised by participants of a normatively oriented organization under attack (whatever the objective circumstance), highly committed members will be most critical. Since staff commitment varies with hierarchical level, these will be persons in the upper reaches of the organization. As such, it is probable that they will perceive that they have an ability to affect agency strategy and will attempt counteraction. This appears to explain the finding that staff who were committed *and* involved in decision-making were more likely to be negative about MFY's crisis stance than those who were committed but less influential.

It also suggests that a strategy of value compromise stemmed less from a long time perspective than it did from lesser commitment, at least in MFY's case. The executives, as the superordinate staff group, might be expected to be more positive about compromise for long range value gains, although, in fact, they were the most critical of all participants. Moreover, although the board of directors included more uncommitted members than any staff group, it also included a larger percentage of negative crisis reactors than either the supervisors or the practitioners. This suggests once again that with influence comes more intense criticism of final decisions which go counter to one's own views.

The dislocating nature of value change is highlighted by the findings. The ideological commitment of high-ranking participants provides the very basis for their ultimate alienation when they perceive value shifts to have occurred.

The anomalous finding of the lesser commitment of MFY's board of directors to the agency's values than MFY's other constituent groups deserves comment. It supports our earlier comment regarding the crudity of the normative organization schema, and provides some evidence that the basis of involvement, particularly of high-ranking participants, significantly affects the degree of their commitment to the values of an organization. A considerable number of Mobilization's board members were involved for utilitarian motives. The initial board group consisted wholly of lay and professional representatives of the settlements and churches of the Lower East Side who hoped to "saturate" the area with services. Their plan for delinquency prevention originally "emphasized group work and recreational services . . . with the settlements as the pivotal community institution" (Piven, 1963:1). Although the board subsequently became more cosmopolitan in its composition, local institutional representation still accounted for a significantly disproportionate number of its active members. The primary concern of this sub-grouping was the funds allocated through contracts by MFY to conduct programs in their agencies.

The primacy of utilitarian incentives for these high-ranking participants allowed lower-ranking normatively oriented actors to define agency values in times of stability. The threat to the organization during the attack—and therefore to their utilitarian rewards—would encourage their defense of the organization as an instrumentality, rather than as a transmitter of values.

In short, the MFY findings suggest, if they do not definitively confirm, that normative organizations with significant variation in the bases of participants' involvement (as is true of most social agencies and many professional organizations) are subject to variations in participants' reactions to the organization's defense strategy. Such organizations are thus particularly vulnerable to the development of internal tensions when threatened by outside conflict.

NOTES

1. For a more complete description of the project, see Brager and Purcell, 1967.

2. Nominally, MFY's board consisted of 68 members, many of whom served symbolic purposes only. Only those who attended more than one meeting during the crisis period are included in this study. Of these 37 persons, one represented New York City, ten were nominated by the Columbia University School of Social Work, five were residents of the community without institutional affiliation, four were board members at large (two from labor and one each from a black and a Puerto Rican organization), and 17 were from local institutions.

3. Sixty-nine percent of MFY's participants disagreed with the statement "MFY's staff was appropriately involved in deciding what strategies to pursue during the attack"; 14% agreed, and 17% had no opinion.

4. Fifty-six percent agreed, as compared to 24.5% who did not, that "MFY's strategies were overly influenced by what City officials would think." Fifty-six percent thought too little consideration was shown for workers who were accused of subversive activities, while 23.2% disagreed.

5. Fifty-three percent felt that "MFY compromised important principles in its handling of the attack"; 27% disagreed.

6. This and other measures are described in the section which follows.

7. In the interests of economy of space, we are summarizing the methods used in the study. For a complete description and explanation of methodology, see Brager, 1967.

8. This use of self-report as an indicator of influence is derived from Gouldner's study of influentials at Co-op College (1957: 297).

9. Consensus, or in any case integration, is also suggested by findings which are not

reported here. For example, only 2% of MFY's personnel reported their jobs to be very or mildly unsatisfying, as compared to a range of 13% to 21% reported in other studies (Herzberg et al., 1957).

REFERENCES

Aubert, Vilhelm
1963 "Competition and dissensus." Journal of Conflict Resolution 7 (March): 26–42.

Boehm, Bernice
1964 "The community and social agency define neglect." Child Welfare 43 (November): 457.

Brager, George
1968 "Organization in crisis: A study of commitment and conflict." New York University, unpublished doctoral dissertation.

Brager, George and Francis Purcell (ed.)
1967 Community Action Against Poverty: Readings from the Mobilization Experience. New Haven: College and University Press.

Coser, Lewis
1961 The Functions of Social Conflict. New York: Free Press.

Etzioni, Amatai
1961 A Comparative Analysis of Complex Organizations. New York: Free Press.
1964 Modern Organizations. Englewood Cliffs: Prentice-Hall.
1965 "Organizational control structure." Pp. 650-677 in James March (ed.), Handbook of Organizations. New York: Rand McNally.

Gouldner, Alvin
1957 "Cosmopolitans and locals." Administrative Science Quarterly II (December): 297.

Herzberg, Frederick, Bernard Mausner, Richard O. Peterson, and Dora Capwell
1957 Job Attitudes: Review of Research and Opinion. Pittsburgh: Psychological Service of Pittsburgh.

Janowitz, Morris
1959 Sociology and the Military Establishment. New York: Russell Sage.

Kramer, Ralph
1965 "Ideology, status, and power in board-executive relationships." Social Work 10 (October): 109.

Mobilization for Youth
1961 A Proposal for the Prevention and Control of Delinquency by Expanding Opportunities. New York: Mobilization for Youth.
1964 Action on the Lower Eastside. New York: Mobilization for Youth.

Piven, Frances
1963 "Conceptual themes in the evolution of mobilization for youth." New York: Mobilization for Youth, mimeographed.

Ross, Aileen
1954 "Philanthropic activity and the business career." Social Forces 32 (March): 274.

Sumner, William Graham
1906 Folkways. Boston: Ginn.

PART IV

Some Consequences of Organizational Structure

The intent of this final section is to demonstrate that the nature of organizations makes a difference. The difference is important for the sociologist interested in organizations. This academic interest is of minor importance when a broader look at organizations in society is taken. Since post-industrial, industrial, and industrializing societies are overwhelmingly organizational societies, the nature of the phenomena under discussion becomes by definition a major consideration at the macro- and micro-levels of societal analysis. Since variations in the structure of organizations and some of the sources thereof have already been demonstrated, attention is now turned to the way in which various organizational forms affect the world around them, the organizational members, and the organizations themselves.

The evidence which will be presented can in no way be viewed as being complete. Indeed, as some radical sociologists have so accurately pointed out, many of the important questions about the impact of the modern organization have hardly been asked, let alone researched. Since we know from sociologists of knowledge that the intellectual style of an era is based on multiple factors, only a small part of which involve pure scientific inquiry and a large part of which involve funding and publication probabilities, this condition is not at all limited to the sociology of organizations. It can be anticipated, hopefully, that attention will increasingly turn to the kinds of issues discussed in this section as the field of sociology itself shifts its focus to more relevant issues and funding and publication possibilities are altered.

While the concern in this section is with the consequences of various organizational forms, the approach which will be taken is within the conceptual framework presented at the outset of this volume. That is, the consequences of organizational forms do not just stop. Rather, there is a continual interaction (feedback) between the organization and the societal elements with which it is in interaction. These interactions thus lead to continual modifications in the organizations themselves. As has been pointed out previously, the exact strength and direction of these relationships is not yet part of our empirical repertoire. If the interest in the implications of organizations for society is in fact translated into research and the research process is not radicalized to the point that it becomes counter-scientific, we may begin to get indications of the relative strength of some of the relationships discussed.

The articles in this section are organized basically around the same conceptual framework as the previous sections in this book. The first paper deals with the impact of the organizational structure on the organization itself. The point here is that particular organizational forms have consequences for the organizations themselves. This in turn will affect how they relate to their environment. Blau's research is an attempt to integrate several different research findings into a set of propositions and corollaries. While Blau's emphasis on size as the major explanatory variable is not consistent with the arguments presented earlier, the findings of this research lead to a vital conclusion for the understanding of organizations. A particular structural arrangement apparently sets in motion a set of processes which continue to transform the organization over time. These processes can be contradictory to the extent that they can counteract one another. This counteraction is not to the point where the organization returns to its original state, but rather involves modifications of trends. In the Blau study, large size and differentiation, which are themselves related, set in motion processes which counteract each other. This kind of research, if extended to a wider variety of organizations and including more variables, will probably yield quite definitive conclusions regarding organizational development. For the purposes of this section it is important to reiterate the point that a major consequence of organizational structuring is the development of altered structural forms. These in turn will lead to further changes. This theoretical point becomes important in practice when the fact that differing organizational forms have consequences for the individuals and societal elements involved in and with the organization is considered.

The next paper deals with the impact of varying organizational structures on the individuals in the organization. Miller pursues further the relationships between a professionalized labor force and the employing organization. His findings are consistent with the general literature on professionals in large organizations—that particular organizational forms are likely to lead to more alienation on the part of this type of employee than other forms. The alienation issue has been approached in other research with essentially the same conclusion.[1] An extension of these findings would be that both alienation and more general forms of conflict and dissatisfaction as well as the more positive rewards received from the employing environment are contingent upon the mix between the organizational system and the characteristics which members bring with them to the organization.[2]

The final set of papers is concerned with the manner in which organizational structure affects transactions with the environment. As noted earlier, this is an area of great interest at present, but one in which relatively little research has been published. The Aiken and Hage paper deals with inter-organizational relationships. It strongly suggests that particular structural arrangements are likely to lead to greater dependence on other organizations. This dependence in turn affects the organizations in an interaction set simply because of the necessity of allocating resources to the interaction. More complex consequences certainly also occur as relationships develop.

The Scott and El-Assal paper represents an attempt to link organizational variables to the student protest movement. The authors find that schools of high quality that are also large and complex in structure are most likely to have experienced student protests.[3] While this to some degree is an issue of the impact of the organization on its members, it rather obviously has a more significant consequence for the wider society.

The kinds of issues raised in these last papers are of growing interest among organizational sociologists. It is clear that insufficient attention has been paid to the consequences of the modern organization for the society surrounding it. While there has been a plethora of studies dealing with alternative management styles and particular roles within the organization, the absence of concern for the wider social system has led to justifiable criticisms that sociologists have not paid sufficient attention to the outputs of organizations as they affect their environment. It is very plausible that different modes of organizing social welfare systems have differential consequences for the clients of the systems. Similarly, it would appear that there are important organizational differences, such as those in control structure, communication patterns, membership composition, and an organization's willingness to attempt to reduce pollution or provide meaningful opportunities for all citizens. In the same vein, it would be interesting to learn if there are organizational variables associated with the power of some organizations over the economy, polity, and total society. While organizational variables would only be part of the answer to these issues, it is a part which has not been dealt with by organizational analysts. Hopefully, this is a condition which will change as research continues.

NOTES

1. For example, see Robert Blauner, *Alienation and Freedom*, Chicago: University of Chicago Press, 1964.

2. In a study of voluntary organizations ("The Forum Theory of Organizational Democracy: Structural Guarantees as Time-Related Variables," *American Sociological Review*, 35–1 [February, 1970]) John Craig and Edward Gross find that the manner in which organizations are structured has distinct implications for the degree of loyalty and extent of participation of members. These are major issues for voluntary organizations and for some aspects of non-voluntary organizations.

3. Gerald Marwell, "Comment on Scott and El Assal" (*American Sociological Review*, 35–5 [October, 1970], p. 916) argues that the Scott-El-Assal data do not really deal with the issue of complexity, but rather demonstrate the overwhelming importance of size. In either case the nature of the organization is the critical factor in regard to the internal protests and external responses.

9

PETER M. BLAU

A FORMAL THEORY OF DIFFERENTIATION IN ORGANIZATIONS

The objective of this paper is to develop a deductive theory of the formal structure of work organizations, that is, organizations deliberately established for explicit purposes and composed of employees. The differentiation of a formal organization into components in terms of several dimensions—spatial, occupational, hierarchical, functional—is considered to constitute the core of its structure. The theory is limited to major antecedents and consequences of structural differentiation. It has been derived from the empirical results of a quantitative study of government bureaus. The extensive analysis of these empirical data on the interrelations between organizational characteristics, too lengthy for presentation in an article, is reported elsewhere (Blau and Schoenherr, 1971).[1] The topic of this paper is not the analysis of the research findings but the deductive theory that can be inferred from them and that therefore explains them and the parallel empirical regularities that the theory predicts to exist in other work organizations. Although the findings are not fully presented, the relevant empirical relationships observed are cited, since they are the basis of the theoretical generalizations advanced, and since they must logically follow from these generalizations to satisfy the requirements of deductive theory.

Reprinted from Peter M. Blau, "A Formal Theory of Differentiation in Organizations," *American Sociological Review*, 35–2 (April, 1970), pp. 201–218.

Deductive Theory

The conception of systematic theory adopted is explicated by Braithwaite (1953; see also Hempel and Oppenheim, 1948; Popper, 1959). An empirical proposition concerning the relationship between two or more variables is explained by subsuming it under a more general proposition from which it can be logically derived. A systematic theory is a set of such logically interrelated propositions, all of which pertain to connections between at least two variables, and the least general of which, but only those, must be empirically demonstrable. "A scientific theory is a deductive system in which observable consequences logically follow from the conjunction of observed facts with the set of fundamental hypotheses in the system" (Braithwaite, 1953:22). The theoretical generalizations that explain the empirical findings are in turn explained by subsuming them under still more general hypotheses, so that the theoretical system may have propositions on several levels of abstraction. These principles apply not only to universal hypotheses—if A, then B—but also to the statistical ones characteristic of the social sciences—the more A, the more likely is B.

The explanatory thrust of a formal theory of this kind resides completely in the generality of the theoretical propositions and in the fact that the empirical findings can be deduced from them in strict logic. Theorizing in the social sciences usually assumes not this form of a deductive model but what Kaplan calls the pattern model, according to which "something is explained when it is so related to a set of other elements that together they constitute a unified system" (1964:333). The psychological experience of gaining understanding by the sudden insight the theory brings of how parts fit neatly into a whole is largely missing in deductive theorizing. Instead, the theorist's aim is to discover a few theoretical generalizations from which many different empirical propositions can be derived. Strange as it may seem, the higher-level hypotheses that explain the lower-level propositions are accepted as valid purely on the basis that they do explain them, in the specific sense that they logically imply them, and without independent empirical evidence; whereas acceptance of the lower-level propositions that need to be explained is contingent on empirical evidence (see Braithwaite, 1953:303). Indeed, the reason for developing a deductive system is to empower empirical findings, confirming low-level hypotheses indirectly to establish

.n abstract body of explanatory theory and empirical evidence for any ɔwer-level proposition strengthens confidence in all propositions.

In Braithwaite's words:

)ne of the main purposes in organizing scientific hypotheses into a deductive ystem is in order that the direct evidence for each lower-level hypothesis may ›ecome indirect evidence for all the other lowest-level hypotheses; although no mount of empirical evidence suffices to prove any of the hypotheses in the ystem, yet any piece of evidence for any part of the system helps toward estab- ;shing the whole of the system (1953: 17–18).

In an attempt to start building a deductive theory of the formal struc- ure of organizations, theoretical generalizations about differentiation in he structure are inferred from a large number of empirical findings of a tudy of government bureaus. Several middle-level propositions are leduced from two basic generalizations, and empirical findings supporting he derived generalizations are cited. Inasmuch as the generalizations ubsume many empirically demonstrated propositions, that is, logically mply them, they explain these empirical regularities. There are several rosswise connections which strengthen the interdependence in the the- retical system.

The aim, in short, is to develop a small number of interrelated general ›ropositions that account for a considerable variety of empirical regulari- ies about differentiation in organizations. The contribution the paper eeks to make rests not on the originality of the particular propositions, everal of which have been noted in the literature, but on the attempt to lerive lower-level propositions systematically from higher-level ones and hus to construct a limited body of coherent theory that is supported by umerous empirical findings. The theory is explicitly confined to infer- nces from the most trustworthy and pronounced empirical relationships ›etween organizational characteristics observed in 1,500 component orga- izations and the 53 larger government agencies to which they belong, in he hope that these strong associations observable under a variety of onditions reflect underlying forces that would also be manifest in other ypes of organizations than the ones studied. A test of most propositions as been conducted in another study of 416 government bureaus of a lifferent kind, but only future research can tell whether and to which xtent the generalizations advanced are also applicable to still other types ›f organizations. Since the theory is restricted to the interdependence mong relatively few factors, it ignores other conditions on which these factors undoubtedly are dependent as well. Thus, the theory pays n attention to the influences of the technology employed, nor to those of th organization's environment. The assumption here is that such other infl ences may complement but do not suppress those of the factors incorp rated in the theory, because these factors are of great general importanc and the empirical data available support this assumption.[2]

Formal Structure

The formal structure of organizations is conceptualized here more na rowly than is usually the case. The term "social structure" is often use broadly, and sometimes loosely, to refer to the common value orientation of people, the traditional institutions in a society, cultural norms and ro expectations, and nearly everything that pertains to life in groups. But has a more specific meaning. The gist of a social structure is that peop differ in status and social affiliation, that they occupy different positio and ranks, and that they belong to different groups and subunits various sorts. The fact that the members of a collectivity are differentiate on the basis of several independent dimensions is the foundation of th collectivity's social structure. This differentiation into components alon various lines in organizations is the object of the present analysis. Th theory centers attention on the social forces that govern the interrelation among differentiated elements in a formal structure and ignores th psychological forces that govern individual behavior. Formal structure exhibit regularities that can be studied in their own right without invest gating the motives of the individuals in organizations.

Formal organizations cope with the difficult problems large-scale oper tions create by subdividing responsibilities in numerous ways and thereb facilitating the work of any operating employee, manager, and subunit i the organization. The division of labor typifies the improvement in pe formance attainable through subdivision. The more completely simpl tasks are separated from various kinds of complex ones, the easier it is fo unskilled employees to perform the routine duties and for skilled emplo ees to acquire the specialized training and experience to perform th different complex ones. Further subdivision of responsibilities occu among functional divisions, enabling each one to concentrate on certai kinds of work. Local branches may be established in different places t facilitate serving clients in various areas, and these branches may becom

unctionally specialized. The management of such a differentiated struc-
ure requires that managerial responsibilities too become subdivided
mong managers and supervisors on different hierarchical levels.

Weber recognized the vital importance the subdivision of responsibili-
ies has for administrative organizations and placed it first in his famous
numeration of the characteristisc of modern bureaucracy.[3] His focus on
structure of differentiated responsibilities is also evident in his emphasis
n the division of labor, specialized competence, and particularly the
ierarchy of authority (see Weber, 1946:196–197; 1947:330–331). An
pparent implication of this stress on structural differences is that the
nalysis of differentiation in the formal structure constitutes the core of
he systematic study of formal organizations, but Weber himself does not
ursue this line of inquiry. It is the primary concern here.

The central concept of differentiation in organizations must be clearly
efined in terms that permit translation into operational measures. A
imension of differentiation is any criterion on the basis of which the
nembers of an organization are formally divided into positions, as illus-
rated by the division of labor; or into ranks, notably managerial levels;
r into subunits, such as local branches, headquarters divisions, or sections
vithin branches or divisions. A structural component is either a distinct
fficial status (for example, employment interviewer or first-line super
isor), or a subunit in the organization (for example, one branch or one
ivision). The term differentiation refers specifically to the number of
tructural components that are formally distinguished in terms of any
ne criterion. The empirical measures used are number of branches, num-
er of occupational positions (division of labor), number of hierarchical
evels, number of divisions, and number of sections within branches or
ivisions.

The research from which the theory or structural differentiation has
een derived is a study of the 53 employment security agencies in the
Jnited States, which are responsible for administering unemployment
nsurance and providing public employment services in the 50 states, the
District of Columbia, Puerto Rico, and the Virgin Islands.[4] These are
utonomous state agencies, although they operate under federal laws and
re subject to some federal supervision. The empirical data were collected
y a team of three research assistants who visited every agency in the
ountry to interview key informants and obtain data from records. Most
f the information about the formal structure comes from personnel lists
nd from elaborate organizational charts specially prepared for the
esearch, all of which were much more detailed than the charts kept by

the agencies. In addition to analyzing the formal structure of the 53 tota agencies or their entire headquarters, the structure of the 1201 loca branches and that of the 354 headquarters divisions were also analyzed these include all local branches and headquarters divisions in the countr meeting minimum criteria of size (five employees) and structure (thre hierarchical levels). Headquarters divisions were, moreover, divided o the basis of their function into six types, making it possible to analyz structure while controlling function. (The six types are the two basi line functions—employment services and unemployment insurance—an four staff functions—administrative services, personnel and technical, dat processing, and legal services.)

First Generalization

Increasing size generates structural differentiation in organizations alon various dimensions at decelerating rates (1). This is the first fundamenta generalization inferred from the empirical findings. From it can b deduced several middle-range propositions, which subsume additiona empirical findings. One can consider this theoretical generalization abou the structure of organizations to comprise three parts, in which case th middle-level and lower-level propositions are derived from the conjunctio of the three highest-level ones. In this alternative formulation, the thre highest-level propositions composing the first basic generalization abou the formal structure of organizations are: (1A) large size promotes struc tural differentiation; (1B) large size promotes differentiation along severa different lines; and (1C) the rate of differentiation declines with expand ing size. The assumption is that these generalizations apply to the subunit within organizations as well as to total organizations, which can be mad explicit in a fourth proposition: (1D) the subunits into which an organiza tion is differentiated become internally differentiated in parallel manne

A considerable number of empirical findings on employment securit agencies can be accounted for by the generalization that differentiatio in organizations increases at decreasing rates with increasing size, an none of the relevant evidence conflicts with this generalization. The opera tional definition of size is number of employees. When total state agencie are compared, increases in size are accompanied by initially rapid an subsequently more gradual increases in the number of local branches int which the agency is spatially differentiated; the number of official occu pational positions expressing the division of labor; the number of levels in the hierarchy; the number of functional divisions at the headquarters; and the number of sections per division. The profound impact that agency size has upon differentiation is indicated by its correlations of .94 with number of local offices; .78 with occupational positions; .60 with hierarchical levels; and .38 with functional divisions. Logarithmic transformation of size further raises these correlations (except the one with local offices); for example, that with number of divisions becomes .54; and that with sections per division, which was before an insignificant .16, is after transformation .43. The improvements in the correlations logarithmic transformation of size achieves reflect the logarithmic shape of the regression lines of the numbers of structural components on size, and thus the declining rate of differentiation with expanding size. For an illustration of this pattern, the scatter diagram for agency size and number of hierarchical levels is presented in Figure 9–1.

The internal differentiation within the subunits that have become differentiated in the agencies assumes the same form. The larger a local branch,

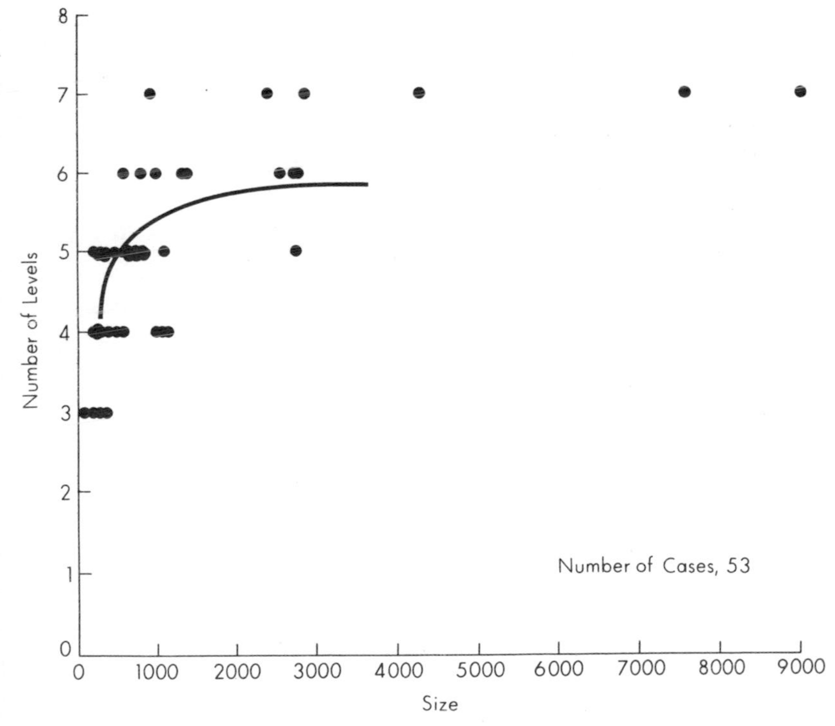

FIGURE 9–1 Size of Agency and Number of Hierarchical Levels at Headquarters.

the greater the differentiation into occupational positions (r = .51), hier archical levels (.68), and functional sections (.61). This differentiation occurs at declining rates with increasing size (and the correlations are somewhat raised when size is logarithmically transformed). The scatter diagram of office size and division of labor (occupational positions) in Figure 9–2 illustrates the logarithmic curve expressing this pattern.[5] Similar logarithmic curves characterize the differentiation within the functional divisions at the agency headquarters. The larger a division, the larger is the number of its occupational positions, hierarchical levels, and functional sections; and differences between very small and medium-sized divisions have again more impact on variations in these three aspects of differentiation than differences between medium-sized and very large divisions. Moreover, this tendency for differentiation at decelerating rates

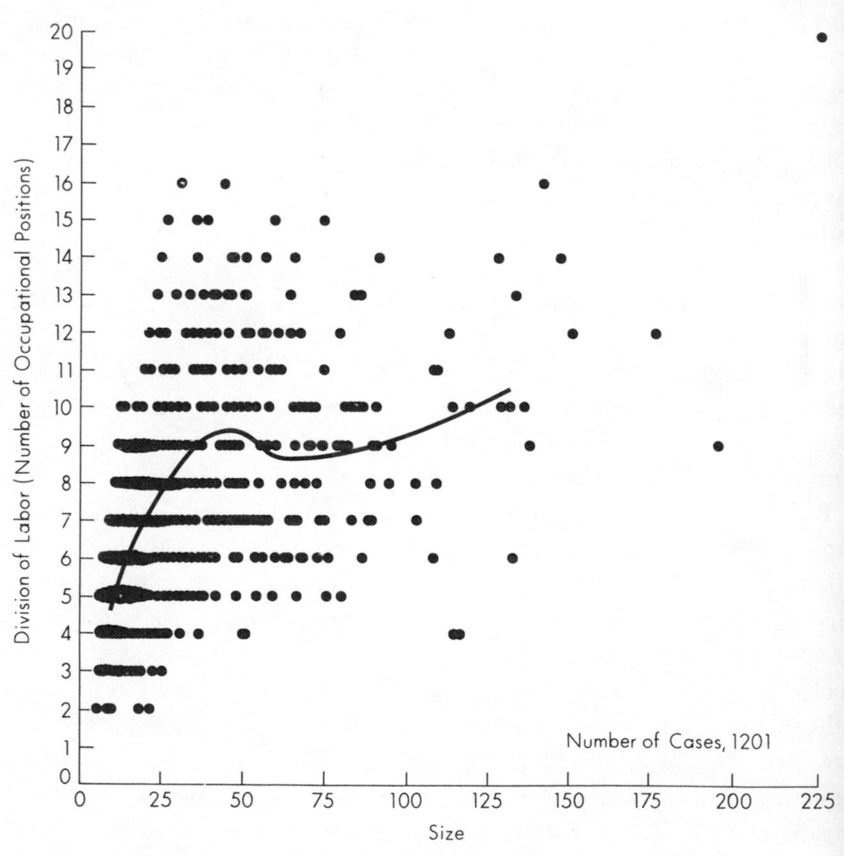

FIGURE 9–2 Size of Local Office and Division of Labor.

o occur with increasing size is observable in six separate types of divi-
ions with basically different functions, which suggests that it is independ-
nt of function and thus provides some support for the claim that the
ame tendency will be found in other organizations which have different
unctions from those of employment security agencies.[6]

PROPOSITION 1.1

The first proposition that can be derived from the first fundamental
eneralization is the following: as the size of organizations increases, its
narginal influence on differentiation decreases (1.1). As a matter of fact,
his is hardly a derived proposition, since it is merely a restatement of
ne part of the original proposition (1C). But by translating the initial
roposition into different concepts, the new proposition directs attention
o a distinctive implication and an important parallel with the economic
rinciple of diminishing returns or, in technical terms, of the eventually
iminishing marginal physical productivity. In the words of Boulding:
'As we increase the quantity of any one input which is combined with a
ixed quantity of the other inputs, the marginal physical productivity of
he variable input must eventually decline" (1955:589).

In a factory, production output can be raised by adding workers, but
he marginal increment in output resulting from adding more and more
vorkers without changing plant size and equipment eventually declines.
n parallel fashion, a larger complement of employees in an organization
nakes its structure more differentiated, but as the number of employees
nd the differentiation of the structure increase, the marginal influence
f a given increase in personnel on further differentiation declines. It
eems that the differentiation produced by the expanding size of organiza-
ions stems the power of additional expansions in size to make the struc-
ure still more differentiated.

But why does the marginal influence of size on differentiation in
rganizations decline? If the analogy with the economic principle of
iminishing returns is appropriate, it should provide some clues for
nswering this question. The reason for the eventually declining marginal
roductivity of increments in only one type of economic input is that
uch increments create an imbalance of inputs and the growing need for
ther inputs depresses productivity. For example, additional workers can-
ot be efficiently utilized in production without parallel increases in
quipment and space. We may speculate that the influence of increasing
rganizational size on differentiation produces a growing need which in
urn diminishes the influences of further increases in size.

The existence of differentiation in a formal organization implies a need for coordination. There are at least two inputs, using the terminology of economics, on which the development of structural differentiation in organization depends. The first is a sufficient number of employees (the measure of size) to fill the different positions and man the various subunits, and the second is an adequate administrative machinery to meet problems of coordination. The advancing differentiation to which an increasing number of employees gives rise intensifies the need for coordination in the organization, and this need restrains the further development of differentiation, which is reflected in the declining marginal influence of increasing size on differentiation. The implication of these considerations extrapolated from economic theory is that differentiation in organizations creates pressures to find ways to meet the need for coordination. We shall later return to the analysis of this problem, after discussing five other propositions that can be derived from the first basic generalization.

PROPOSITION 1.2

The second derived proposition is that the larger an organization is, the larger the average size of its structural components of all kinds (1.2). This proposition logically follows from the principle of the decelerating rate of differentiation with increasing organizational size (1C), which is graphically expressed by the decline in the slopes (the logarithmic curves) of the regression lines of the number of structural components of various kinds on size. In a diagram with total size or number of employees on the horizontal and number of structural components on the vertical axis, the average size of these components necessarily becomes larger. Even in point is indicated by the ratio of the horizontal to the vertical coordinate of this point. As the positive slope of the regression line declines, this ratio is larger for most large agencies than for most small agencies.[7] If the structural components increase more slowly than organizational size, the average size of these components necessarily becomes larger. Even in those cases in which the decline in the slope of the regression line is not pronounced, the average size of the structural components is strongly associated with the size of the organization. Two examples are the mean size of local branches, which variable has a zero-order correlation with agency size of .65, and the number of incumbents of the average occupational position, which variable is correlated .94 with agency size.

Thus, the large size of an organization raises the average size as well as the number of its structural components. Large agencies have more and larger local offices than small agencies, more and larger headquarters divisions, and the same hold true for every one of their structural components. The large size of the local offices within an agency and of its headquarters divisions, whatever their function, in turn tends to increase both the number and the average size of sections, and both the number of occupational positions and of managerial ranks and the mean number of employees occupying each position and each rank.

This double effect of organizational size has the paradoxical result that large offices and headquarters divisions constitute at the same time a more homogeneous and a more heterogeneous occupational environment for most employees than small ones. For larger offices or divisions contain comparatively many employees in nearly every occupational specialty, providing a congenial ingroup of colleagues for most employees—often not available in small organizational units—and they simultaneously contain a relatively great variety of different specialties, enhancing opportunities for stimulating contacts with people whose training and experience are unlike their own. However, the greater opportunity for social interaction with a colleague ingroup in large offices may prove so attractive that social contacts with persons from different specialties are rarer there than in small ones, despite the fact that opportunities for outgroup contacts are better in large offices too.

PROPOSITION 1.3

A third derived proposition is that the proportionate size of the average structural component, as distinguished from its absolute size, decreases with increases in organizational size (1.3). This follows directly from (1A): if the number of structural components, the criterion of differentiation, increases as organizational size does, the proportion of all employees who are in the average component must decrease. Hence, most groups or categories of employees in big organizations are larger in absolute numbers but constitute a smaller proportion of the total personnel than in small organizations. A consequence is that the average (*mean*) relative size of employee complements on a given dimension decreases with increasing organizational size, though not necessarily the proportion of any particular complement.

But we may reformulate this proposition (1.3) into a probability statement about groupings of employees: *ceteris paribus*, chance expectations are that the proportionate size of any personnel complement decreases with increasing organizational size. The empirical data show that this proposition applies to various kinds of administrative overhead or supportive services for the majority work force. The size of an agency is inversely related to the proportionate size of its administrative staff (r=—.60) and of its complement of managerial personnel (—.45). (The terms "manager" and "supervisor," unless qualified, are used interchangeably to refer to all levels.) The proportion of managers is also inversely related to size in local offices (—.64) and in headquarters divisions regardless of function.[8]

When a certain personnel complement is singled out for attention—the staff or the managerial component—and exhibits the expected decrease in proportionate size with increasing organizational size, the remainder of the total personnel—the line or the nonsupervisory employees—must naturally reveal a complementary increase in proportionate size. This is mathematically inevitable, and it indicates that the reformulated proposition (1.3) cannot possibly apply to both parts of a dichotomy. The plausible assumption is that the residual majority actually consists of numerous personnel categories while the specialized personnel complement focused upon can be treated as a single one, which implies that the proportion of the minority complement is the one that should decrease with increasing organizational size. The data support this assumption. If employees in various organizational units are divided into clerical and professional personnel, the proportion of whichever of the two is in the minority tends to decrease as unit size does. The conclusion that may be drawn, which extends beyond what can be derived in strict logic from the premise, is that the proportionate size of any supportive service provided by a distinctive minority to the majority work force is likely to decline with increasing organizational size.

PROPOSITION 1.4

Another proposition can be derived either from the last one (1.3) or from the one preceding (1.2): the larger the organization is, the wider the supervisory span of control (1.4). If chances are that the proportionate size of any organizational component declines with increasing size (1.3), and if this applies to the proportion of managers, it follows that the number of subordinates per manager, or the span of control, must expand with increasing size (1.4). Besides, if chance expectations are that the absolute average size of any structural component or grouping of employees increases with increasing size (1.2), and if this applies to the various work groups assigned to supervisors, it follows that the size of the group under each supervisor, or his span of control, tends to expand with increasing size (1.4). Here again the logical implications specifying the *mean* absolute and proportionate size for *all* components have been translated into *probabilities* or statistical expectations referring to *any* component. Whether these derived propositions apply to a *particular* personnel component, like the managerial staff, must be empirically ascertained. If the evidence is negative, it would not falsify the theory, though it would weaken it. If the evidence is positive, it strengthens the theory, and makes it possible to extend it beyond the limits of its purely logical implications by taking into account the empirical data confirming this particular application of the merely statistical deduction from the theory.[9]

The empirical data on employment security agencies confirm the proposition that the span of control of supervisors expands with increasing organizational size. This is the case for all levels of managers and supervisors examined in these agencies and their subunits. The larger an agency, the wider is the span of control of its director and the average span of control of its division heads. The larger a headquarters division, whatever its function, the wider is the span of control of its division heads, the average span of control of its middle managers, and the average span of control of its first-line supervisors. The larger a local office is, the wider the span of control of the office manager and that of the average first-line supervisor.[10] Moreover, the size of the total organization has an independent effect widening the supervisory span of control when the size of local offices is controlled.[11] Big organizations and their larger headquarters divisions and local branches tend to have more employees in any given position with similar duties than small organizations with their smaller subunits, as we have seen, thus making it possible to use superviors more efficiently in large units by assigning more subordinates with similar duties to each supervisor.

The additional influence of the size of the total organization, independent of that of the size of the office, on the number of subordinates per supervisor, may reveal a structural effect (see Blau, 1960). The prevalence of a wide span of supervisory control in large organizations, owing to the large size of most of their branch offices, creates a normative standard that exerts an influence in its own right, increasing the number of subordinates assigned to supervisors; and the same is the case, *mutatis mutandis*, for the prevalence of a narrow supervisory span of control in small organizations with their smaller branches. To direct attention to the substantial influence of organizational size on the supervisory span of control is, of course, not to deny that this span is also influenced by other conditions, such as the nature of the duties.

PROPOSITIONS 1.5 AND 1.6

Organizations exhibit an economy of scale in management. This proposition (1.5) is implicit in the two foregoing ones. For if the proportion of managerial personnel declines with size (1.3) and their span of control expands with size (1.4), this means that large-scale operations reduce the proportionate size of the administrative overhead, specifically, of the complement of managers and supervisors. In fact, the relative size of administrative overhead of other kinds, such as staff and supportive personnel, also declines with increasing size, as has been noted. The question arises whether this economy of scale in administrative overhead produces overall personnel economies with an increasing scale of operations. The data on employment security agencies are equivocal on this point. The only index of personnel economy available, the ratio of all employees engaged in unemployment benefit operations to the number of clients served by them, is inversely correlated with size, but with a case base of only 53 agencies the correlation is too small (—.14) to place any confidence in it. Logarithmic transformation of size raises the correlation to —.24, which suggests that large size might reduce the man-hour costs of benefit operations slightly.

Whereas this finding is inconclusive, not inconclusive are the numerous findings that indicate that the relative size of administrative overhead declines with increasing organizational size. Large-scale operations make it possible to realize economies in managerial manpower. This can be explained in terms of the generalization that the number of structural components increases at a declining rate with increasing size (1), which implies that the *size* of work groups under a supervisor, just as that of most personnel components, increases with increasing size, and that the *proportion* of supervisors, just as that of most personnel components, decreases with increasing size, and these relationships account for the economy of scale in management.

A final derived proposition in this set is that the economy of scale in administrative overhead itself declines with increasing organizational size (1.6). This proposition follows from two parts of the basic generalization (1A and 1C) in conjunction with one derived proposition (1.3). If the number of structural components increases with increasing organizational size (1A), the statistical expectation is that the proportionate size of any particular personnel component decreases with size (1.3). The empirical data showed that the proportion of managerial personnel and that of staff personnel do in fact decrease as size increases, in accordance with these expectations. But since the increase in the number of components with expanding size occurs at a declining rate (1C), the decrease in the proportionate size of the average component, implicit in this increase in number, must also occur at a declining rate with expanding organizational size. Reformulation in terms of statistical probability yields the proposition that chance expectations are that the proportionate size of any particular personnel complement decreases at a decelerating rate as organizations become larger.

Whether this statistical proposition about most personnel components holds true for the managerial and the staff component is an empirical question, and the answer is that it does. The proportion of staff personnel decreases at a declining rate as organizational size increases (see Figure 9–3), and so does the proportion of managerial personnel at the agency headquarters as well as in local offices (see Figure 9–4). The marginal power of organizational size to produce economies in administrative overhead diminishes with growing size, just as its marginal power to generate structural differentiation does. Both of these patterns are implied by the generalization that the number of structural components in an organization increases at a declining rate with expanding size.

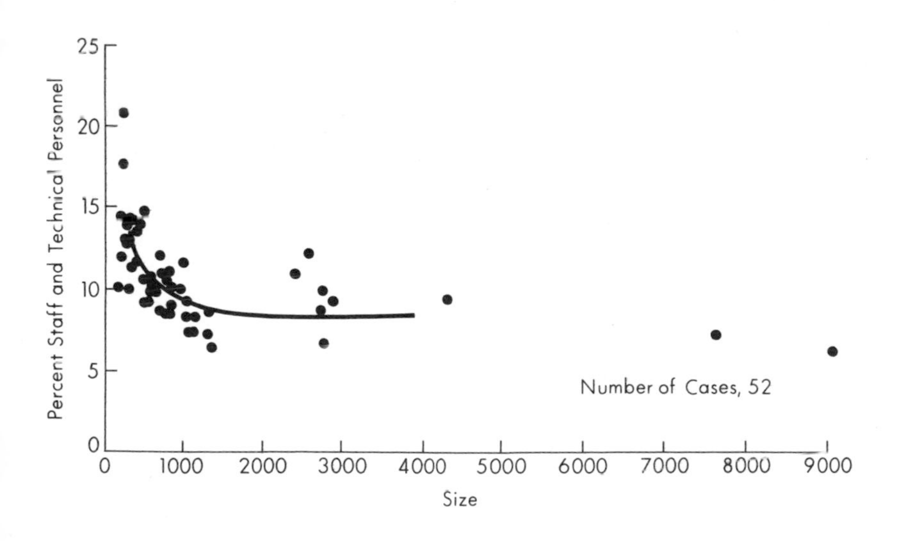

FIGURE 9–3 Size of Agency and Percent Staff Personnel.

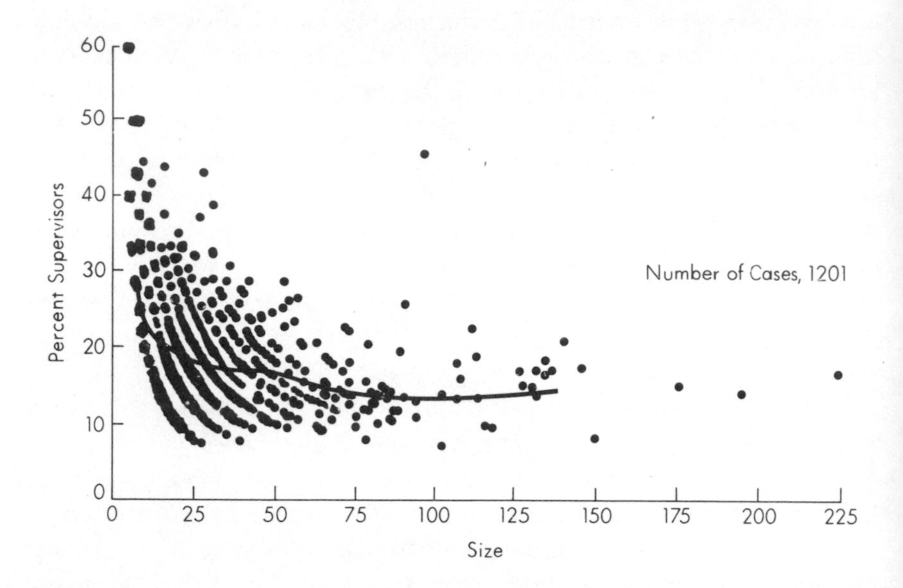

FIGURE 9–4 Size of Local Office and Percent Supervisory Personnel.

Transition

The structure of formal organizations seems to undergo repeated social fission with growth. In a large organization, its broad responsibilities tend to be subdivided to facilitate their performance, and it thereby becomes differentiated into a number of structural components of diverse sorts. The larger an organization, however, the larger is typically not only the number but also the average size of the components into which it is differentiated. These larger segments of larger organizations, in turn, tend to become internally differentiated along various lines. Thus, the process of social fission recurs within the differentiated units which that process produced. Differentiation lessens the difficulties the performance of duties entails by reducing the scope of the responsibilities assigned to any individual or unit, but it simultaneously enhances the complexity of the structure. Social fission makes duties less complex at the expense of greater structural complexity.

When responsibilities become extensively subdivided, many employees will have the same duties and entire units will have similar ones, and savings in supervisory manpower may occur. At the same time, however, the greater structural complexity implicit in the pronounced subdivision of large organizations intensifies problems of communication and coordination, which make new demands on the time of managers and supervisors at all levels. In short, the very differentiation of responsibilities through which large organizations facilitate the performance of duties and reduce the need for supervision creates fresh administrative problems for supervisory personnel. The theory presented so far accounts for the effect of size on savings in supervisory manpower (1.4, 1.5) as well as for its effect on differentiation (1A), but it does not yet include an explicit proposition concerning the effect of differentiation on supervision and administration.

To be sure, the analysis of the proposition that the marginal influence of an organization's size on differentiation declines with increasing size (1.1) has already led to the inference that differentiation intensifies administrative problems. The assumption is that the problems of coordination and communication in differentiated structures have feedback effects that create resistance to further differentiation, which is the reason why the marginal influence of size on differentiation declines with increasing size. The expanding size of an organization is a social force that produces differentiation. The more differentiated an organization is, according to this interpretation, the more resistance a force must overcome to produce still more differentiation, and the more of an expansion in size it therefore takes to effect a given increment in differentiation.

This interpretation seeks to explain the decelerating rate at which increasing size generates differentiation in organizations, but it cannot be logically deduced from the propositions referring to this decelerating rate. It is important in this connection to keep in mind the distinction between inferring a higher-order generalization from a lower-order proposition, in an inductive argument, and logically deriving a lower-order from a higher-order proposition, in deductive reasoning. What is logically implied by the generalization that the rate of differentiation declines with expanding size (1C), as well as by the derived proposition that the marginal influence of increasing size on differentiation diminishes (1.1), is that differentiation gives rise to *some* problems and needs that stifle the further development of differentiation, as indicated by the decreasing power of size to effect differentiation. It does not follow, though it is a plausible inference, that these are problems of coordination and communication calling for administrative solutions. Hence, another basic generalization is postulated incorporating these ideas, which explains some of the propositions in the first set, and which in conjunction with earlier propositions yields three more derived propositions.

Second Generalization

Structural differentiation in organizations enlarges the administrative component (2), because the intensified problems of coordination and communication in differentiated structures demand administrative attention. In this second fundamental generalization of the deductive theory, the first part subsumes many empirical findings, whereas the second part introduces theoretical terms not independently measured in the research but inferred. The assumptions are that differentiation makes an organization more complex; that a complex structure engenders problems of communication and coordination; that these problems create resistance to further differentiation; that managers, the staff, and even first-line supervisors spend time dealing with these problems; and that consequently more supervisory and administrative manpower is needed in highly differentiated structures than in less differentiated structures. Although these assumptions of the intervening connections are not empirically tested, the implications of the conclusion are. If, in accordance with the inferred assumptions, much of the time of supervisors on all levels in the most differentiated structures is occupied with problems of communication and coordination, it follows that these supervisors have less time left for guiding and reviewing the work of subordinates.

Hence, the more differentiated the formal structure, the more administrative personnel of all kinds should be found in an organization of a given size, and the narrower the span of control of first-line supervisors as well as higher managers. This is precisely the pattern the empirical findings reveal. Vertical differentiation into levels and horizontal differentiation into divisions or sections are both positively related to the proportion of supervisors among the total personnel, controlling size, in the whole organization, in local branches, and in the six functional types of headquarters divisions. They are also positively related to the proportionate size of the staff in agencies of a given size.

Moreover, both vertical and horizontal differentiation, with size held constant, are negatively related to the span of control of managers and supervisors on different levels in local offices and in headquarters divisions, regardless of function.[12] The finding that the second generalization and its derivations discussed below are supported when the span of control of supervisors on a given level is substituted for the ratio of all supervisors is of special importance. The more levels organizations of a given size have, the larger is necessarily the proportion of their supervisors, that is, of their personnel above the lowest level. The positive relationship of number of levels with proportion of supervisors does not merely reflect this mathematical nexus, which would make it trivial, as demonstrated by its positive relationship with supervisory span of control, which is not affected by this nexus. Hence, the empirical data support the principle that hierarchical as well as horizontal differentiation, presumably by engendering problems of coordination, enlarges requirements for managerial manpower.

PROPOSITIONS 2.1 AND 2.2

One derived proposition is that the large size of an organization indirectly raises the ratio of administrative personnel through the structural differentiation it generates (2.1). If increasing organizational size generates differentiation (1A), and if differentiation increases the administrative component (2), it follows that the indirect effect of size must be to increase the administrative component. Decomposition of the zero-order correlations of size with various ratios of managerial and staff personnel in multiple regression analysis makes it possible to isolate the indirect effects of size mediated by differentiation from its direct effects. In every problem analyzed, the empirical findings confirm the prediction that the indirect effects of size mediated by both vertical differentiation into levels and horizontal differentiation into divisions or sections raise the ratio of administrative to total personnel. This is the case whether the dependent variable under consideration is the staff ratio or the managerial ratio at the agency headquarters; the ratio of supervisors on all levels; or the span of control of first-line supervisors in any of the six types of functional divisions or in local branches. In all these instances, the indirect effects of size mediated by the differentiation it generates and its direct effects are in opposite directions. The savings in administrative overhead large-scale operations make possible are counteracted by the expansion in administrative overhead the structural complexity of large organizations necessitates.

Another derived proposition is that the direct effects of large organizational size lowering the administrative ratio exceed its indirect effects raising it owing to the structural differentiation it generates (2.2). This is a logical consequence of propositions (1.5) and (2.1). If the overall effect of large size reduces management overhead (1.5), and if large size, by fostering differentiation, indirectly increases management overhead (2.1), it follows that its effect of reducing overhead must outweigh this indirect effect. All the decompositions of the zero-order correlations of size with various measures of management reflect this, as they inevitably must. For example, the direct effect of agency size on the managerial ratio at the agency headquarters, which is represented by the standarized regression coefficient when three measures of differentiation are controlled, is —1.13, whereas its overall effect, indicated by the zero-order correlation, is —.45, the difference being due to the strong counteracting effect mediated by differentiation.[13] For the staff ratio at the agency, with the same conditions controlled, the direct effect of size is —1.04, and its overall effect is —.60, revealing again a substantial indirect counteracting effect due to structural differentiation. The direct and indirect effect of the size of a division on its managerial ratio and of the size of a local office on its managerial ratio reveal parallel differences.[14] *Ceteris paribus*, a large scale of operations would effect tremendous savings in administrative overhead, but these savings are much reduced by the structural differentiation of large organizations. Consistently, however, the economies of scale exceed the costs of differentiation, so that large organizations, despite their greater structural complexity, require proportionately less administrative manpower than small ones.

PROPOSITION 2.3

The last proposition to be derived is that the differentiation of large organizations into subunits stems the decline in the economy of scale in management with increasing size, that is, the decline in the decrease in the proportion of managerial personnel with increasing size (2.3). The derivation of this proposition is rather complicated and must be approached in several steps. The new proposition is not as well knit into the system as the others and should be regarded as a mere conjecture.

The concept of economy of scale in administration refers to the fact that the proportion of various kinds of administrative personnel decreases with the increasing size of the organization or its subunits. The operational indication is a negative correlation between any of these proportions and size, which is represented on a graph by a negative slope of the regression line of the proportion on size. These negative correlations and slopes are evident in all empirical data on employment security agencies: size of local branch and either proportion of all managerial personnel or ratio of first-line supervisors to operating employees (the reverse of span of control); size of functional division and either ratio of all managerial personnel or ratio of supervisors to subordinates on three levels; size of total agency and either proportion of staff personnel, or proportion of managerial personnel at the headquarters, or proportion of managerial personnel in the total organization.

A decline in this economy of scale means that the *rate of decrease* in the ratio of managerial personnel itself *decreases* with increasing size. This is reflected on a graph by a curve in the negative slope of the regression line of the ratio on size that shows that the ratio of overhead personnel drops first sharply and then more gradually with increasing size. The per cent of supervisors in local offices illustrates a decrease at such a decreasing rate (Figure 9–4), and so does the ratio of staff personnel in the agency (Figure 9–3) and that of the supervisors at the agency headquarters (not shown), and the same pattern is observable in most other relationships mentioned in the above paragraph. The major exception is that the proportion of managerial personnel in the total agency does not reveal such a declining rate of decrease but a fairly linear decrease with increasing agency size, as Figure 9–5 shows. Although this appears to be a deviant case, the principle it expresses can be deduced from the propositions in the theory.

In local offices, the smallest organizational unit examined, the proportion of all supervisory personnel drops rapidly as size increases from ten, or fewer, to about fifty employees, but it drops much more slowly with further increases to one and two hundred employees (see Figure 9–4).

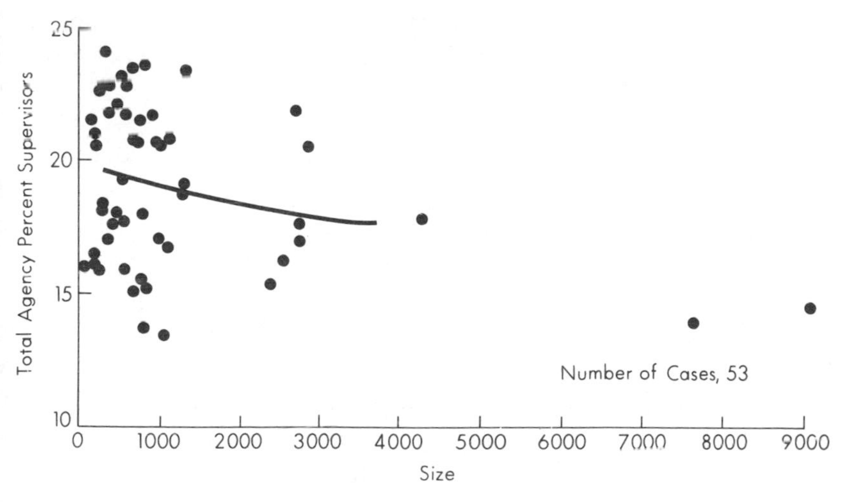

FIGURE 9–5 Size of Agency and Percent Supervisory Personnel in Total Agency.

From a projection of this trend, one would expect that further increases in size to several thousand employees are hardly accompanied by any decline in the proportion of supervisory personnel. As the size of the entire organization increases from about one hundred to several thousand employees, however, the total proportion of supervisory personnel decreases on the average at a constant rather than declining rate, as Figure 9–5 reveals, though there is much scatter. Although this decline is not pronounced, it is by no means inconsequential; the zero-order correlation is —.34, which compares with a correlation of —.46 between size of office and its proportion of supervisors. (However, the latter correlation is raised to —.64 if size is logarithmically transformed. In contrast, the former correlation is reduced to —.23 by such a transformation, which is another indication that the regression line does not exhibit a logarithmic curve.) Why does the decrease in the proportion of managerial personnel with increasing size, which is already very small as office size expands beyond fifty employees, not become virtually zero but is again considerable as agency size expands from several hundred to several thousand? The answer suggested by the theory is that the differentiation of large organizations into many branch offices (and divisions), while raising the proportion of managers needed, simultaneously restores the economy of scale in the managerial component, that is, it recreates the decline in the proportion of managerial personnel with increasing size observed among very small organizational units.

The growing need for managerial manpower resulting from the structural differentiation engendered by expanding size (2.1) increasingly impinges upon the savings in managerial manpower that a large scale of operations realizes (1.5), which helps to explain why the economy of scale in management declines as size and differentiation increase (1.6).[15] In other words, the *rate* of savings in management overhead with increasing size is higher among comparatively small than among comparatively large organizational units, although, or perhaps because, the management overhead is bigger in small than in large organizational units. Differentiation in a large organization (1A) means that it consists of relatively many smaller rather than relatively few larger organizational subunits, such as local offices. Inasmuch as the *rate* of savings in management overhead is higher in smaller than in larger organizational units, the reduction in the size of units created by differentiation raises this rate of savings and stems the decline in the economy of scale with respect to management overhead that would be otherwise expected once organizations have grown beyond a certain size (2.3).

Conclusions

A formal theory of the formal structure of formal organizations has been presented. Its subject is formally established organizations with paid employees, not emergent social systems or voluntary associations of people. It is confined to the analysis of the formal structure—specifically, its differentiation—of organizations, ignoring the informal relations and behavior of individuals, within these organizations. And the endeavor has been to develop a formal theory by inferring from many empirical findings a minimum number of generalizations that can logically account for these findings. These findings come from a quantitative study of all employment security agencies and their subunits in the United States. The two basic generalizations, from which nine other propositions were deduced, are: (1) the increasing size of organizations generates structural differentiation along various dimensions at decelerating rates; and (2) structural differentiation enlarges the administrative component in organizations.

The concluding review of the theory rearranges the order of presentation of propositions to call attention to alternative connections between them and to some of the unmeasured terms assumed to underlie these connections. Organizing the work of men means subdividing it into component elements. In a formal organization, explicit procedures exist for systematically subdividing the work necessary to achieve its objectives. Different tasks are assigned to different positions; specialized functions are allocated to various divisions and sections; branches may be created in dispersed locations; administrative responsibilities are subdivided among staff personnel and managers on various hierarchical levels. The larger an organization and the scope of its responsibilities, the more pronounced is its differentiation along these lines (1A, 1B), and the same is the case for its subunits (1D). But large-scale operations, despite the greater subdivision of tasks than in small scale operations—involve a larger volume of most organizational tasks. Hence, large organizations tend to have larger as well as more structural components of various sorts than small organizations (1.2).

The pronounced differentiation of responsibilities in large organizations enhances simultaneously intra-unit homogeneity and inter-unit heterogeneity. Inasmuch as duties are more differentiated and the amount of work required in most specialities is greater in large organizations than in small ones, there are comparatively many employees performing homogeneous tasks in large organizations. The large homogeneous personnel components in large organizations simplify supervision and administration, which is reflected in a wider span of control of supervisors (1.4) and a lower administrative ratio (1.3) in large than in small organizations. Consequently, organizations exhibit an economy of scale in administrative manpower (1.5). At the same time, however, the heterogeneity among organizational components produced by differentiation creates problems of coordination and pressures to expand the administrative personnel to meet these problems (2). In this formulation, the unmeasured concepts of intra-unit homogeneity and inter-unit heterogeneity have been introduced to explain why large size has two opposite effects on administrative overhead, reducing it owing to the enlarged scale of similar tasks, and raising it owing to the differentiation among parts.

By generating differentiation, then, large size indirectly raises administrative overhead (2.1), and if its influence on differentiation were unrestrained, large organizations might well have disproportionately large administrative machineries, in accordance with the bureaucratic stereotype. However, the administrative ratio decreases with expanding organizational size, notwithstanding the increased administrative ratio resulting from the differentiation in large organizations (2.2). Two feedback effects of the administrative costs of differentiation may be inferred, which counteract the influences of size on administration and differentiation, respectively. The first of these apparently reduces the savings in administrative manpower resulting from a large scale of operations, as implied by the decline in the rate of decrease of administrative overhead with increasing organizational size (1.6). (Although differentiation into local branches may keep the rate of overhead savings with increasing size constant [2.3], it also raises the amount of overhead.) The second feedback process, probably attributable to the administrative problems engendered by differentiation, creates resistance to further differentiation which is reflected in the diminishing marginal influence of expanding size on differentiation (1.1) and the declining rate at which size promotes differentiation (1C).

In short, feedback processes seem to keep the amount of differentiation produced by increasing organizational size below the level at which the additional administrative costs or coordination would equal the administrative savings realized by the larger scale of operations. Hence, organizations exhibit an economy of scale in administration, despite the extra administrative overhead required by the pronounced differentiation in large organizations, but this economy of scale declines with increasing size on account of this extra overhead due to differentiation. The feedback effects inferred, though not directly observable, can explain why the influence of size on differentiation, as well as its influence on administrative economy, declines with increasing size. Figure 9–6 presents these connections graphically.

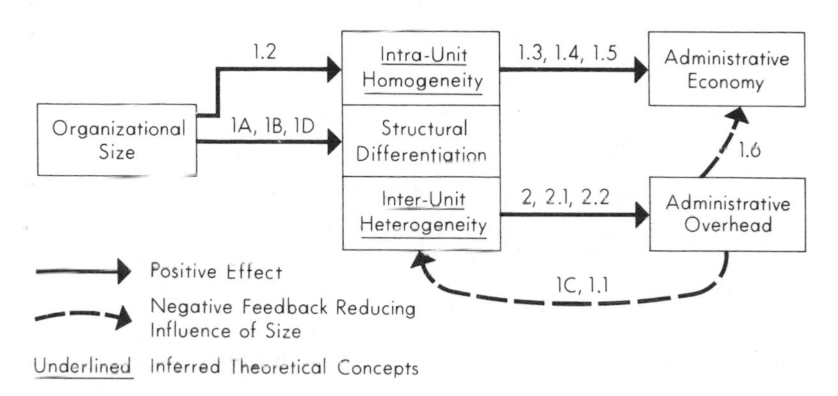

FIGURE 9–6 Chart of Connections.

A final question to be raised is how widely applicable the theory is to organizations of different types. Since the theory was constructed by trying to formulate generalizations from which the empirical findings on employment security agencies can be derived, the fact that these data conform to the propositions advanced does not constitute a test of the theory. But it should be noted that several of the specific propositions included in the theory are supported by findings from previous empirical studies of other kinds of organizations, for example, that administrative overhead in organizations decreases with size and increases with complexity or differentiation (see Anderson and Warkov, 1961; Pondy, 1969, and references therein), and that its decrease with size occurs at a declining rate (Indik, 1964). Moreover, an empirical test of the entire body of theory has been conducted in a study of another type of government bureau, the 416 major finance departments of American states, large cities, and large counties. This independent test confirms the propositions implied by the theory.[16] Whether the theoretical generalizations are also valid for private and other public organizations and how they must be modified or refined to make them widely applicable, only further research can tell.

NOTES

1. The assistance of Sheila R. Klatzky with this research is gratefully acknowledged, and so is grant GS–553 of the National Science Foundation, the source of support of

the Comparative Organization Research Program at the University of Chicago, of whic this is report No. 11.

2. The research of Blau and Schoenherr (*The Structure of Organizations*, New York Basic Books, 1971) presents data that show that the empirical relationships implied b the theory persist when important differences in technology and numerous variations i environmental conditions are controlled.

3. "I. There is the principle of fixed and official jurisdictional areas . . . 1. The regu lar activities required for the purposes of the bureaucratically governed structure ar distributed in a fixed way as official duties" (Max Weber, 1946: 196).

4. The only agency excluded is the smallest one, on Guam, which has less than dozen employees, compared to 1,200 for the mean of the other agencies. In the fou jurisdictions in which unemployment insurance and employment services are carried ou by separate bureaus, they were combined for the purpose of the analysis, since it becam evident that these two functions are hardly more separate there than in some other juris dictions where they are legally in the same bureau.

5. The curves shown are rough estimates. They were drawn by dividing size int three categories for Figure 9–1 and seven categories for Figure 9–2; determining th means for both size and the y-variable (ordinate) in each category; and making a smoot curve between those points. The same procedure is used for the other figures below.

6. This pattern is reflected in the finding that the logarithmic transformation of siz improves its zero-order correlations, for all types of divisions combined, with division o labor (from .64 to .76), hierarchical levels (from .71 to .85), and functional section (from .38 to .68). If the six functional types are analyzed separately, logarithmic trans formation of size raises the corresponding correlations in 15 of 18 cases.

7. This can be readily seen by looking at the regression lines in Figures 9–1 and 9–2 For a point moving along either line from left to right, the horizontal coordinate increase more rapidly than the vertical one, indicating that the ratio of the first to the secon coordinate increases.

8. The zero-order correlations for the six types of divisions are: —.49 (employmen services); —.51 (unemployment insurance); —.30 (administrative services); —.12 (per sonnel and technical); —.18 (data processing); and —.36 (legal services). Size in all case (agencies and local offices as well as divisions) has been logarithmically transformed

9. Two kinds of statistical or probability statements must be distinguished, empiric and theoretical ones. On the one hand, it is only probable that any given large agenc has a lower ratio of supervisors than any given small agency, since the correlation is les than 1.00; this empirical probability is *not* what is referred to in the text. On the othe hand, and this is what is discussed above, it is only probable that the ratio of super visory personnel is inversely related to agency size, since the theory only predicts tha the proportionate size of most components of the agency is inversely related to its ow size and that it is probable that such an inverse relationship will be observed with respec to any particular component, such as the supervisory ratio.

10. The zero-order correlations of size (log) of the respective organizational unit and mean span of control of various managers are: agency director, .39; head of divisio from .22 to .44 for the six functional types; middle managers in divisions, with on exception (.05) from .17 to .78; first-line supervisors in divisions, from .39 to .69 i the six types; managers of local offices, .40; and first-line supervisors in local offices, .66

11. In the multiple regression problem with the average span of control of the first line supervisors in each of the 1201 local offices as dependent variable, and with offic size and a number of other conditions controlled, the standardized regression coefficie of the size of the agency to which the local office belongs is .27.

12. This statement and those in the preceding paragraph are based on a multipl regression analyses with size (log) and a number of other conditions controlled; tw

or three measures of differentiation as the independent variables (levels, divisions, and sections per division in agencies; levels and sections in local offices and in divisions); and the following dependent variables: for agencies, managerial ratio and staff ratio; for local offices, managerial ratio, span of control of office manager, and mean span of control of first-line supervisors; for the six types of divisions, managerial ratio, span of control of division head, mean span of control of middle managers, and mean span of control of first-line supervisors.

13. The three aspects of differentiation controlled in this problem, as well as in the one mentioned in the next sentence, are number of (a) levels, (b) divisions, and (c) sections per division.

14. In the multiple regression analysis for all divisions combined (with sections, levels, clerical ratio, division of labor, agency size, and agency managerial ratio controlled), the standardized regression coefficient indicating the direct effect of a division's size (log) on its managerial ratio is —1.32, and the zero-order correlation indicative of the overall effect is only —.23, with differentiation into levels (.65) and sections (.35) being responsible for most of the difference. The separate regression analyses for the six types yield parallel results. In the analysis of local offices (with levels, sections, specialization, manager's span of control, and division of labor controlled), the standardized regression coefficient of office size (log) on the managerial ratio is —1.43, but this incredibly strong direct effect is reduced to a still substantial overall effect, represented by the zero-order correlation, of —.64, most of the reduction being due to differentiation into levels (.41) and sections (.40).

15. This alternative derivation of proposition (1.6) illustrates the type of crosswise connections that creates a more closely knit theoretical system. Other alternative connections are presented in the conclusions.

16. The number of occupational positions, that of hierarchical levels, and that of functional divisions in these finance departments increases at decelerating rates with increasing size, as indicated by regression lines with positive slopes and logarithmic curves (1, 1.1, 1.2, 1.3). The span of control of first-line supervisors is positively correlated with size (1.4), which implies that the supervisory ratio decreases with size (1.5), and this relationship reveals a logarithmic curve, indicating that the economy of scale in supervisory manpower declines with size (1.6). The numbers of levels and that of divisions both raise the ratio of supervisors to non-supervisory personnel (2), and large size indirectly does so through its influences on levels and divisions (2.1), but these indirect effects of size are exceeded by its direct effect reducing the ratio of supervisors by widening their span of control (2.2). The absence of local branches in finance departments makes it impossible to test proposition (2.3).

REFERENCES

Anderson, Theodore R. and Seymour Warkov

1961 "Organizational size and functional complexity." American Sociological Review 26: 23–28.

Blau, Peter M.

1960 "Structural effects." American Sociological Review 25: 178–193.

Blau, Peter M. and Richard A. Schoenherr

1971 The Structure of Organizations. New York: Basic Books.

Boulding, Kenneth E.

1955 Economic Analysis (3d ed.). New York: Harper.

Braithwaite, Richard B.
1953 Scientific Explanation. Cambridge, England: Cambridge University Press.
Hempel, Carl G. and Paul Oppenheim
1948 "The logic of explanation." Philosophy of Science 15: 135–175.
Indik, Bernard P.
1964 "The relationship between organization size and supervision ratio." Administrative Science Quarterly 9: 301–312.
Kaplan, Abraham
1964 The Conduct of Inquiry. San Francisco: Chandler.
Pondy, Louis R.
1969 "Effects of size, complexity, and ownership on administrative intensity." Administrative Science Quarterly 14: 47–60.
Popper, Karl R.
1959 The Logic of Scientific Discovery. New York: Basic Books.
Weber, Max
1946 Essays in Sociology. New York: Oxford University Press.
1947 The Theory of Social and Economic Organization. New York: Oxford University Press.

10

GEORGE A. MILLER

PROFESSIONALS IN BUREAUCRACY: ALIENATION AMONG INDUSTRIAL SCIENTISTS AND ENGINEERS*

Etzioni has said that most of us are born in organizations, educated by organizations, work for organizations, and spend much of our leisure time paying, playing, and praying in organizations.[1]

The modern scientist and engineer represent dramatic examples of the incorporation of professionals into organizations. At present, nearly three-fourths of all scientists in the United States are employed by industry and government, with the greatest proportion of these in industry.[2] Previous research findings show that businessmen are increasingly dependent upon the scientist and engineer for ideas necessary for the accomplishment of their organizational goals[3] and that the amount of research conducted by staff scientists in industrial organizations is an important factor differentiating the dynamic from the more stagnant industries.[4] The employment of scientists and engineers in industry is not increasing at the same rate in all types of industry—as evidenced by the increase in the numbers of these professionals who are employed in the aerospace industry.[5] Moreover, the aerospace industry itself has become more and more dependent

Reprinted from George A. Miller, "Professionals in Bureaucracy: Alienation among Industrial Scientists and Engineers," American Sociological Review, 32–5 (October 1967), pp. 755–768.

* This investigation was supported in part by a Public Health Service Fellowship from the National Institutes of Health. The author is greatly indebted to L. Wesley Wager, University of Washington, for aid and criticism in the formulation of this research and to Samuel J. Surace, Warren D. TenHouten, and Ralph H. Turner for their many helpful comments on an earlier draft of this paper.

upon the federal government as an increasing proportion of its research and development is supported directly by the National Aeronautics and Space Administration and the Department of Defense.[6]

This growing interdependence has created conflicts for both the professional and his employing organization. Blau and Scott have shown that, although professional and bureaucratic modes of organization share some principles in common, they rest upon fundamentally conflicting principles as well.[7] Kornhauser concludes that most conflicts between the scientist or engineer and his employing organization stem from the basic organizational dilemma of autonomy vs. integration. These professionals must be given enough autonomy to enable them to fulfill their professional needs, yet their activity must also contribute to the overall goals of the organization.[8]

The professional who experiences such conflicts in his work may become alienated from his work, the organization, or both. This research examines work alienation as one major consequence arising from the professional-bureaucratic dilemma for industrial scientists and engineers.

Although it is clear that conflicts exist between professionals and their employing organizations, it is also clear that organizational administrators are not unaware of the problems and conflicts inherent in the employment of professional personnel. The research to date suggests that these organizations are attempting to become more professional and flexible. One method for alleviating conflicts is to modify the organizational structure by providing more *professional incentives* and lessening the degree of *organizational control* for the professional employee.[9]

However, these changes in the organizational structure may not be made available to all professionals. This paper will suggest that structural modification in professional incentives and organizational control will vary by (1) the length and type of training received by the professional, and (2) the type of organizational unit in which the professional performs his work activity.

PROFESSIONAL TRAINING

Orth describes the importance of professional training for the scientist or engineer who enters a bureaucratic organization, as follows:

Professional training in itself, whether it be in medicine, chemistry, or engineering, appears to predispose those who go through it to unhappiness or rebellion when faced with the administrative process as it exists in most organizations. Scientists and engineers *cannot* or *will not* . . . operate at the peak of their creative potential in an atmosphere that puts pressure on them to conform to organizational requirements which they do not understand or believe necessary.[10]

Kornhauser contends that the strength of professional loyalty and identification can be expected to vary by the *type* of training received by the professional. He notes that professions differ in their selectivity of recruitment, intensity of training, and state of intellectual development and that all of these factors affect the nature of the person's orientation.[11] Becker and Carper in a study of three groups of students found differences between those majoring in engineering and those majoring in physiology and psychology. The engineers felt that their future lay somewhere in the industrial system. For many, a broad range of positions within an organization was thought acceptable, including such "unprofessional" positions as manager or research supervisor.[12] Similarly, Clovis Shepherd found differences between scientists and engineers concerning their goals, reference groups and supervisory experiences. Engineers were typically more "bureaucratic" in their orientation, whereas scientists were more "professional." Shepherd attributes this difference to the more intensive training received by the scientists.[13] Goldner and Ritti also contend that engineers are more concerned with entrance into positions of power and participation in the affairs of the organization than with professional values and goals. They argue that engineers recognize power as an "essential ingredient of success" for them.[14]

In addition to differences in type of professional training, differences in the *length* of training are important. Those professionals who receive the Ph.D. degree should develop stronger professional loyalties and identifications than those persons with M.A. or M.S. degrees. This greater length of training also represents a greater investment on the part of the professional. It is reasonable to assume that he will *expect* higher rewards from the organization in return for his services. Moreover, if these rewards are not forthcoming, professionals with advanced degrees are in a better position to market their skills and find employment in an organization that does offer such rewards.

Some evidence exists for these assumptions. For example, Wilensky found that top staff experts in labor unions who had received serious job offers since their employment had more influence on union decisions than did those without such offers.[15] Paula Brown found that "key" scientists and engineers in a governmental laboratory were given "special considerations" by their administrators, were highly respected by them, and were able to "control" the laboratory to a considerable extent. Brown argues that professionals with less national prestige would not have been given the same treatment nor have been able to control the organization as effectively.[16] Pelz's and Andrews' findings parallel those reported earlier in a study undertaken by the Princeton Opinion Research Corporation. In both studies, those professionals with the Ph.D. degree and those trained as scientists participated most often in decisions regarding their work, had more individual freedom, and enjoyed more professional incentives.[17]

However, the "match" between organizational structure and the professional's length and type of training is seldom perfect. As Etzioni notes:

> If personalities could be shaped to fit specific organizational roles, or organizational roles to fit specific personalities, many of the pressures to displace goals, much of the need to control performance, and a good part of the alienation would disappear. Such matching is, of course, as likely as an economy without scarcity and hence without prices.[18]

The above line of reasoning serves as the basis for the following hypotheses to be explored in this paper: (1) degree of alienation from work should be positively associated with degree of organizational control and negatively associated with number of professional incentives for all professional personnel; (2) the above relationships should be stronger for those professionals with the Ph.D. degree and those professionals trained as scientists than for professionals with the M.A. or M.S. degree and professionals trained as engineers.

ORGANIZATIONAL UNIT

An equally important conditioning variable is the type of organization in which the professional performs his work. Comparative organizational research shows that the greater the organization's dependence upon research, the more the organization will manifest high professional incentives and low organizational control.[19]

Structural variations *within* an organization may also be apparent. One instance where structural variation might be expected in an organization employing scientists and engineers is between those areas where application and development are the primary goals and those areas where the primary goal is pure or basic research. Most aerospace companies are engaged in both types of activity, but the basic research aspect is usually carried out in a special laboratory that is separated from the larger organizational unit.

This research allows for a comparison of these two very different types of organizational units and thus permits a further evaluation of the hypothesized relationship between organizational structure and work alienation. Specifically, there should be less organizational control and more professional incentives in the laboratory than in the larger unit. Therefore, less alienation from work should be experienced by professionals employed in the laboratory than by professionals employed in the larger organizational unit.

Method

Data were gathered during the summer of 1965 from scientific and engineering personnel employed in two divisions of one of the largest aerospace companies in the United States. The organization, at the time of the study, employed over 100,000 persons within its five operating divisions and had separate facilities in four states and subsidiaries in Canada. The company is engaged in the manufacture of military aircraft, commercial airliners, supersonic jet transports, helicopters, gas-turbine engines, rocket boosters for spacecraft, intercontinental ballistic missiles, and orbiting space vehicles.

The largest of the five divisions is the Aero-Space Group, containing a missile production center and a space center. Included within the Aero-Space Group are a variety of independent laboratories, testing facilities, manufacturing areas, and the world's largest privately owned wind tunnel. Shortly after this study was undertaken, a large space-simulation laboratory was completed.

In 1958 the company established a Basic Science Research Laboratory which operates independently from all other laboratories and divisions of the company. The principal product of the Laboratory is the scientific and engineering information made known to other divisions of the company by means of consultation and publication programs. The laboratory has three major objectives: (1) to carry on research programs that will put the company into direct and effective communication with the larger scientific community; (2) to choose and maintain a staff of competent specialists who will provide consulting advice in their chosen fields of specialization; and (3) to engage in exploratory and basic research in fields where new discoveries will be of value to the company's overall operations.[20]

The two divisions thus represent a sharp contrast in the nature of the work situation for the professionals involved. Professionals in the Basic Science Laboratory share an environment more like that of the university, whereas professionals in the Aero-Space Group are more representative of persons engaged in traditional research and development work and function primarily as staff personnel within the division.

Subjects for this study are non-supervisory scientists and engineers selected from the Aero-Space Group and the Basic Science Laboratory. All subjects held the degree of M.A., M.S., or Ph.D. in science, engineering, or mathematics. A listing of all employees meeting the above criteria was provided by the company and the selection of respondents was made from this listing. Twenty different types of engineers were represented, ranging from aerospace to electrical to nuclear engineers. Scientists included astronomers, chemists, and physicists. General, applied and theoretical mathematicians were included. Respondents within the two divisions were selected by the following criteria:

1. All Basic Science Laboratory personnel (N=66),
2. All persons in the Aero-Space Group with the degree of Ph.D. (N=74),
3. All persons in the Aero-Space Group with the degree of M.A. or M.S. in science or mathematics (N=164),
4. A 50 percent random sample of all persons in the Aero-Space Group with the degree of M.A. or M.S. in engineering (N=236).

In the analysis to follow, those persons with degrees in mathematics are included with those having degrees in science and both are labeled "scientists." The engineering personnel are treated as a separate group of "engineers." Because of the sampling design, the engineering group is under-represented in terms of their actual proportion in the organization although their proportion in the sample is greater than for any other professional specialty.

Data were gathered by means of a mailback questionnaire sent to the homes of the study participants. The questionnaire was anonymous and a postcard was included for the respondents to sign and return after they had returned their questionnaires. This procedure allowed for subsequent follow-ups in the case of non-respondents and two such follow-ups were undertaken.

Seventeen of the original 540 subjects had either moved from the area or the addresses provided by the company were incorrect, leaving 523 potential subjects. Of this total, 84 percent returned the questionnaires and 80 percent (N = 419) were completed sufficiently to be used in this analysis.

Measures

WORK ALIENATION

The work of professionals, as contrasted with the work of non-professionals, is characterized by high intrinsic satisfaction, positive involvement, and commitment to a reference group composed of other professionals.[21] In addition, as Orzack has demonstrated, the work of professionals plays a much more important role in their life than it does for the non-professional worker.[22]

For a scientist, the importance of and devotion to a particular style of life and work was best described by Weber. In his essay concerning science as a vocation, he indicates the importance of "inward calling" for the scientist:

> . . . whoever lacks the capacity to put on blinders, so to speak, and to come up to the idea that the fate of his soul depends upon whether or not he makes the correct conjecture at this passage of his manuscript may as well stay away from science. He will never have what one may call the "personal experience" of science. Without this strange intoxication, ridiculed by every outsider; without this passion, this "thousands of years must pass before you enter into life and thousands more wait in silence"—according to whether or not you succeed in making this conjecture; without this, you have *no* calling for science and you should do something else. For nothing is worthy of man as man unless he can pursue it with passionate devotion.[23]

The scientist who is unable to find "self-rewarding work activities to engage him," who does not experience an "intrinsic pride or meaning in his work," and who "works merely for his salary" or other remunerative incentives is experiencing the type of alienation described by Melvin Seeman as self-estrangement.[24] As defined, self-estrangement is *not* the same as dissatisfaction with one's job. As Mills and others have pointed out, a person may be alienated from his work yet still be satisfied with his job.[25]

The measure of work alienation employed is a five-item cumulative scale consisting of statements referring to the intrinsic pride or meaning of work. Three statements were developed by the author and two were selected from Morse's scale of "intrinsic pride in work."[26] The statements are:

1. I really don't feel a sense of pride or accomplishment as a result of the type of work that I do.
2. My work gives me a feeling of pride in having done the job well.
3. I very much like the type of work that I am doing.
4. My job gives me a chance to do the things that I do best.
5. My work is my most rewarding experience.

Response categories provided for each statement were: (1) Strongly Agree; (2) Agree; (3) Disagree; and (4) Strongly Disagree. The response distribution to each item was dichotomized between those agreeing and those disagreeing with the statement. This procedure yielded a Guttman scale with the following characteristics: Coefficient of Reproducibility (Goodenough technique) = 0.91; Minimum Marginal Reproducibility = 0.70; Coefficient of Scalability = 0.69; and Coefficient of Sharpness = 0.69.[27]

The scale scores were trichotomized to obtain three levels of work alienation. Persons with a scale score of "0" (30 percent) were classified as low, those with scale scores of "1" or "2" (40 percent) were classified as medium, and those with scale scores of "3," "4," or "5" (30 percent) were classified as high in experiencing alienation from work.[28]

ORGANIZATIONAL CONTROL

Organization implies the coordination of diverse activities necessary for effective goal achievement. Such coordination requires some mode of control over these diverse activities. Traditional bureaucratic modes of control differ significantly from the mode of control deemed appropriate by professionals. Whereas organizations tend to be structured hierarchically, professions tend to be organized in terms of a "colleague group of equals" with ultimate control being exercised by the group itself.[29] Hence, bureaucratic control violates the profession's traditional mandate of freedom from control by outsiders.[30]

One specific area where the nature of the organizational control structure becomes manifest is in the type of supervisor-employee relationship. This is a major source of conflict for the professional as it is for other types of organizational participants. In this respect, Baumgartel suggests that three styles of supervision (or leadership) can be identified, based upon rates of interaction, degree of influence, and decision-making. The three styles of supervision are: *Directive* (low rate of interaction and unilateral decision-making by the supervisor); *Participatory* (high rate of interaction and joint decision-making by supervisor and researcher); and *Laissez-Faire* (low rate of interaction, with the researcher making most of the decisions).[31]

In his study of scientists and supervisors in the National Institutes of Health, Baumgartel found that research performance, job satisfaction, and positive attitudes toward the supervisor were highest for Participatory, lowest for Directive, and intermediate for Laissez-Faire supervisors. These findings differ sharply from those reported by Argyris. Eighty-seven percent of the foremen studied by Argyris report that in order to be effective, they must try to keep everyone busy with work that guarantees a fair takehome pay, to distribute the easy and tough jobs fairly, and *to leave the employees alone as much as possible*. Argyris concludes that a successful foreman, from the point of view of both foremen and employees, is neither directive nor is he an expert in human relations.[32]

To determine the types of supervision evident in the present organization, the following question was asked:

Which of the following statements most nearly represents the type of work relationship that exists between you and your immediate supervisor?

1. We discuss things a great deal and come to a mutual decision regarding the task at hand.
2. We discuss things a great deal and his decision is usually adopted.
3. We discuss things a great deal and my decision is usually adopted.
4. We don't discuss things very much and his decision is usually adopted.
5. We don't discuss things very much and I make most of the decisions.

Persons responding with statement number 4 above were classified as working for a Directive type supervisor (16 percent), those responding with statement number 5 above were classified as working for a Laissez-Faire type (45 percent), and those responding with statements 1, 2, or 3 above were classified as working for a Participatory type supervisor (39 percent).

The second empirical indicator of organizational control concerns the professional's freedom to choose or select the types of research projects in which he is implicated. Leo Meltzer found that scientists in industrial organizations usually have ample funds and facilities for research but very little freedom in the selection of research projects. Conversely, scientists in universities are usually short on funds but have freedom to do what they wish. Both freedom and funds were found to be correlated with publication rate, but were negatively correlated with each other. Meltzer concludes that freedom to choose research projects is representative of other variables which facilitate the intrinsic satisfactions which scientists derive from the actual content of their work.[33]

To ascertain the degree of freedom accorded professionals in this study, the following question was asked:

In general, how much choice do you have concerning the types of research projects in which you are involved?

1 ... Almost no choice 3 ... Some
2 ... Very little 4 ... A great deal

Persons responding "Almost no choice" or "Very little" were classified as low (38 percent), those indicating they had "Some" choice were classified as medium (40 percent), and those indicating they had "A great deal" of choice were classified as high (22 percent) in freedom to choose research projects.

PROFESSIONAL INCENTIVES

Previous research findings describe the incentives most sought after by scientists and engineers as (1) freedom to publish the results of their research, (2) funds for attending professional meetings, (3) freedom and facilities to aid in their research, (4) promotion based upon technical competence, and (5) opportunities to improve their professional knowledge and skills.[34] It is more common, however, for industry to slight professional incentives in favor of organizational incentives, such as promotions in the line, increases in authority, and increases in salary. This is the case because organizational incentives have proved satisfactory for other employees in the past and differing incentive structures are viewed as competing sources of loyalty.[35]

Two empirical indicators of the professional incentive structure are utilized in this research. The first, *professional climate*, is a general index of the professional incentives made available to the professional by the organization and is composed of two items:

1. (Company) provides us with many opportunities to obtain professional recognition outside the company.

1 ... Strongly disagree 3 ... Agree
2 ... Disagree 4 ... Strongly agree

2. In general, how much time are you provided to work on or pursue your own research interests?

1 . . . Almost none 3 . . . Some
2 . . . Very little 4 . . . A great deal

The response distribution for each item was dichotomized between responses 2 and 3. Persons agreeing with the first statement and persons responding "some" or "a great deal" to the second question were classified as high. These two dichotomies were then cross-classified to obtain three levels of perceived professional climate: those high on both dimensions (28 percent), those high on one dimension (34 percent), and those low on both dimensions (38 percent).

A second indicator, *company encouragement*, attempts to discern the organization's encouragement of specific professional incentives. Three items comprise this index:

1. (Company) encourages us to publish the results of our research.
2. (Company) encourages us to attend our professional meetings.
3. (Company) encourages us to further our professional training by attending special lectures and/or classes at academic institutions.

Response categories provided for each question were Strongly Disagree, Disagree, Agree, and Strongly Agree. The response distribution for each item was dichotomized between those who agreed and those who disagreed with the statement. Those disagreeing with all or two of the statements were classified as low (35 percent), those agreeing with two of the statements were classified as medium (28 percent), and those agreeing with all three statements (37 percent) were classified as perceiving high encouragement from the company.

Results

The first hypothesis states that alienation from work will be positively associated with degree of organizational control and negatively associated with number of professional incentives. Table 10–1 shows support for this general hypothesis.

There are fairly large differences between those professionals working for Directive and Participatory supervisors in the proportions with differing degrees of alienation. However the proportions of those with differing

TABLE 10-1
Relationship between Work Alienation and Four Indicators of Organizational Control and Professional Incentive Structures

Control-incentive structure	*Work alienation*			
	Low	*Med.*	*High*	*N*
Supervisor type				
Directive	.10	.33	.57	63
Participatory	.32	.42	.26	158
Laissez-faire	.35	.42	.23	176
		Gamma = −.30		
Research choice				
Low	.11	.34	.55	150
Med.	.32	.48	.20	158
High	.56	.40	.04	87
		Gamma = −.64		
Professional climate				
Low	.15	.34	.51	155
Med.	.29	.45	.26	133
High	.51	.43	.06	106
		Gamma = −.55		
Company encouragement				
Low	.20	.39	.41	142
Med.	.26	.38	.36	107
High	.42	.44	.14	144
		Gamma = −.35		

degrees of work alienation who work for Participatory and Laissez-Faire supervisors are almost identical. Thus, the major difference in the degree of alienation experienced by these professionals occurs between those who work for a Directive Supervisor and those working for *either* of the other types.

The relationship between company encouragement and work alienation is much the same. Differences exist between those professionals with low and high company encouragement who experience alienation; however, the proportions of those experiencing differing degrees of alienation with medium and low company encouragement are very similar.

LENGTH OF PROFESSIONAL TRAINING

The original relationships just examined are expected to be conditional relationships. It was hypothesized that the relationships should be stronger for those professionals with advanced training. Table 10–2 shows strong support for this expectation.

The relationship obtained in each partial is consistent in direction with that expected and the degree of association is much stronger for those professionals with the Ph.D. degree. In addition, the form of the relationships involving type of supervisor and company encouragement differs in the two partials.

The importance of length of professional training on the relationship between type of supervisor and work alienation is clear. The relationship obtained for those with the M.A. degree is similar to the original relationship. For those with the Ph.D. degree, however, type of supervisor is important and is reflected in the differences in degree of work alienation evident among professionals working for the three types of supervisors. The proportions with various degrees of alienation are about the same for both Ph.D.'s and M.A.'s who work for a Directive supervisor. However, the Ph.D.'s differ from the M.A.'s with respect to the importance of the other two supervisor types. While there is no difference among those M.A.'s who experience high alienation, there is a 17 percent difference among the Ph.D.'s. Conversely, for those experiencing low alienation there

TABLE 10.2
Relationship between Work Alienation and Four Indicators of Organizational Control and Professional Incentive Structures: By Length of Professional Training

Control-incentive structure	Length of professional training: M.A. degree, Work alienation: Low	Med.	High	N	Ph.D. degree, Work alienation: Low	Med.	High	N
Supervisory type								
Directive	.10	.30	.60	52	.11	.33	.56	9
Participatory	.28	.47	.25	124	.46	.24	.30	33
Laissez-faire	.27	.47	.26	122	.56	.31	.13	52
	Gamma = −.25				Gamma = −.40			
Research choice								
Low	.11	.36	.53	131	.17	.17	.66	18
Med.	.33	.50	.17	127	.33	.34	.33	27
High	.42	.53	.05	38	.67	.31	.02	49
	Gamma = −.59				Gamma = −.71			
Professional climate								
Low	.17	.34	.49	132	.09	.32	.59	22
Med.	.26	.50	.24	107	.48	.22	.30	23
High	.38	.55	.07	58	.67	.29	.04	48
	Gamma = −.45				Gamma = −.68			
Company encouragement								
Low	.18	.42	.40	117	.29	.29	.42	24
Med.	.21	.42	.37	79	.41	.22	.37	27
High	.34	.48	.18	100	.63	.32	.05	41
	Gamma = −.29				Gamma = −.48			

is a 10 percent difference between those working for the two supervisor types. In fact, 56 percent of those with the Ph.D. degree experience low alienation under a Laissez-Faire supervisor as compared with only 27 percent of those with the M.A. degree.

The relationships obtained between company encouragement and work alienation are similar. As with type of supervisor, the partial obtained for those with the M.A. degree is very similar to that observed in the original relationship, with little difference in degree of alienation evident for those with low and medium company encouragement. However, for the Ph.D.'s, these differences are more apparent and the degree of association is very different in the two partials. For the Ph.D.'s, the difference between those perceiving low and high company encouragement who experience high work alienation is 37 percent as compared with only 12 percent for those with the M.A. degree. Conversely, the same difference for those experiencing low alienation is 34 percent for the Ph.D.'s and only 14 percent for the M.A.'s.

TYPE OF PROFESSIONAL TRAINING

The strength of the original relationships is also expected to vary with type of professional training. Specifically, the relationships should be stronger for those trained as scientists than for those trained as engineers. Table 10–3 shows only partial support for this expectation.

The relationship between type of supervisor and work alienation again remains consistent in direction but differs greatly in magnitude between the two partials. In the partial obtained for the engineering personnel, the relationship is similar to that observed in the original relationship and among professionals with the M.A. degree. Working for a Directive supervisor is associated with high alienation but little difference is evident in the degree of alienation between those persons working for Participatory and Laissez-Faire supervisors. In the partial obtained for the scientific personnel, however, there is clearly a difference in the degree of alienation experienced by professionals working for all three types of supervisors. Working for a Directive supervisor leads to high alienation, but 11 percent more scientists experience high alienation under Participatory supervisors than under Laissez-Faire supervisors (as compared with a difference of only 4 percent for the engineering group).

Thus, the relationship between type of supervisor and degree of work alienation is conditional with respect to type and length of professional training. For those professionals with M.A. degrees or who have been trained as engineers, the major differences in alienation appear between

TABLE 10-3
Relationship between Work Alienation and Four Indicators of Organizational Control and Professional Incentive Structures: By Type of Professional Training

Control-incentive structure	*Type of professional training*							
	Engineers Work alienation				*Scientists Work alienation*			
	Low	*Med.*	*High*	*N*	*Low*	*Med.*	*High*	*N*
Supervisory type								
Directive	.09	.35	.56	32	.10	.28	.62	29
Participatory	.30	.42	.28	90	.34	.42	.24	67
Laissez-faire	.21	.47	.32	89	.51	.36	.13	85
	Gamma = −.12				Gamma = −.49			
Research choice								
Low	.10	.38	.52	92	.14	.28	.58	57
Med.	.26	.52	.22	89	.43	.40	.17	65
High	.59	.37	.03	27	.55	.42	.03	60
	Gamma = −.61				Gamma = −.60			
Professional climate								
Low	.16	.31	.53	89	.15	.37	.48	65
Med.	.20	.53	.27	74	.43	.34	.23	56
High	.41	.50	.09	46	.59	.38	.03	60
	Gamma = −.49				Gamma = −.58			
Company encouragement								
Low	.21	.42	.37	97	.19	.35	.46	54
Med.	.23	.35	.42	60	.30	.40	.30	46
High	.24	.53	.23	62	.57	.45	.08	79
	Gamma = −.13				Gamma = −.55			

those working for a Directive as opposed to *either* a Participatory or Laissez-Faire supervisor. For those persons with the Ph.D. degree or who were trained as scientists, however, the less the supervision (as all three types are differentiated), the less the degree of alienation from work.

The relationship between company encouragement and work alienation is consistent in direction with that obtained in the original relationship but is much stronger for scientists than for engineers. For the scientists, the difference between those who perceive high and low company encouragement who experience high alienation is 38 percent, as compared with a difference of only 14 percent for the engineers. Conversely, there is little difference in degree of alienation among engineers who perceive low as opposed to high company encouragement. For the scientists, however, there is a corresponding difference of 38 percent.

Thus, the relationship between company encouragement and alienation from work is consistent with the general hypotheses. The greater the company encouragement, the less the feelings of work alienation among professionals. This relationship is stronger for those with the Ph.D. degree and for those trained as scientists than for those with the M.A. degree and for those trained as engineers.

Type of professional training *does not* appear to be an important conditioning factor in the relationships involving freedom of research choice or professional climate. The relationships obtained are strong but similar in both partials. These two variables were most strongly associated with work alienation in the original relationships. The degree of association remained high in the partials when length of professional training was controlled (although the relationships were stronger for the Ph.D.'s than for the M.A.'s, as expected). Unlike the relationships obtained with type of supervisor or company encouragement, however, the relationship between these variables and work alienation remain very strong and about the same for both scientific and engineering personnel.

This unanticipated finding suggests that these two aspects of the organizational structure are important to *both* scientists and engineers. Moreover, since these relationships also remained strong when length of professional training was controlled, it may be concluded that having freedom to choose research projects and working in a professional atmosphere are more important than type of supervisor or amount of company encouragement as these aspects of the organizational structure are related to the experiencing of work alienation among professionals.

ORGANIZATIONAL UNIT

As previously described, the goals of the Laboratory are more concerned with pure or basic scientific research whereas the goals of the Aero-Space Group are more concerned with traditional research and development work. Therefore, the Laboratory should be characterized by less organizational control and more professional incentives than the Aero-Space Group. In addition, since alienation from work is related to differences in organizational structure, professionals working in the Laboratory should experience less alienation than professionals working in the Aero-Space Group.

Table 10–4 shows striking differences in the control and incentive structures of the two organizational units. In each of the four comparisons, there is less control and more incentives in the Laboratory than in the Aero-Space Group.

That these differences in organizational structure are reflected in differences in the degree of work alienation experienced by the professionals is shown in Table 10–5. In the Laboratory, only 4 percent experienced high

TABLE 10-4
Respondents' Perceptions of Organizational Control and Professional Incentive Structures: By Organizational Unit

Control-incentive structure	*Organizational unit*			
	Aero-Space Group		*Basic Laboratory*	
	Percent	*Number*	*Percent*	*Number*
Supervisor				
Directive	17	(64)	2	(1)
Participatory	43	(158)	20	(10)
Laissez-faire	40	(145)	78	(38)
Research choice				
Low	42	(151)	8	(4)
Med.	44	(162)	6	(3)
High	14	(51)	86	(43)
Professional climate				
Low	44	(158)	0	(0)
Med.	38	(136)	10	(5)
High	18	(64)	90	(45)
Company encouragement				
Low	40	(142)	4	(2)
Med.	28	(100)	20	(10)
High	32	(112)	76	(37)
Total	100	(369)	100	(50)

work alienation as compared with 34 percent of those persons in the Aero-Space Group. Moreover, 63 percent of those in the Laboratory experienced low alienation as compared to only 25 percent of those in the Aero-Space Group. Thus, type of organizational unit is clearly related to degree of work alienation in the expected direction.

A question may arise concerning this interpretation. It is clear that the properties of the organizational structure dealt with here are defined empirically in terms of the professional's *perception* of the type of structure in which he is employed. Therefore, a question may arise as to

TABLE 10-5
Degree of Work Alienation in Aero-Space and Basic Laboratory Organizational Units

Organizational unit	*Work alienation*			
	Low	*Med.*	*High*	*N*
Aero-Space Group	.25	.41	.34	349
Basic Laboratory	.63	.33	.04	49
		Gamma = −.69		

whether the respondents are *accurately* perceiving the structure or merely perceiving it in such a manner as to be *consistent* with their previous training.

This question is a crucial one since all of the previous findings rely upon empirical indicators which ask the respondent to describe the structure as he perceives it. Thus, these relationships may reflect an association between the way these persons perceive the structure and work alienation rather than an association between organizational structure and work alienation. Moreover, it is clear from inspection of the previous tables that there is a "hidden" association between length and type of professional training and work alienation. Therefore, it is possible that the differences in degree of work alienation observed between the two organizational units is simply a result of the differing *composition* of the members within each unit with respect to length and type of professional training. To answer this question, it is necessary to examine the relationship between organizational unit and degree of work alienation while controlling for length and type of professional training.[36]

Tables 10–6 and 10–7 show the partials obtained when length and type of professional training are controlled. It will be noted that the direction of association remains consistent across the four partials. The difference between Aero-Space Group and Basic Laboratory personnel in experiencing low work alienation is over 30 percent in all four partials, with Laboratory personnel experiencing much less alienation than Aero-Space personnel. Moreover, the difference between personnel with high work alienation in the two units is also 30 percent in three of four partials. This finding, then, lends additional support to the findings concerning the effects of organizational structure on feelings of work alienation among professional personnel.

TABLE 10-6
Degree of Work Alienation in Aero-Space and Basic Laboratory Organizational Units: By Length of Professional Training

	Length of professional training							
	M.A. degree Work alienation				*Ph.D. degree Work alienation*			
Organizational unit	*Low*	*Med.*	*High*	*N*	*Low*	*Med.*	*High*	*N*
Aero-Space Group	.23	.44	.33	288	.36	.29	.35	59
Basic Laboratory	.54	.46	.00	13	.67	.28	.05	36
		Gamma = −.48				Gamma = −.59		

TABLE 10-7
Degree of Work Alienation in Aero-Space and Basic Laboratory Organizational Units: By Type of Professional Training

	Type of professional training							
	Engineers Work alienation				*Scientists Work alienation*			
Organizational unit	*Low*	*Med.*	*High*	*N*	*Low*	*Med.*	*High*	*N*
Aero-Space Group	.21	.43	.36	199	.31	.38	.31	147
Basic Laboratory	.58	.42	.00	12	.65	.30	.05	37
	Gamma = –.75				Gamma = –.62			

Summary and Conclusions

This study examined the relationship between the type of organizational structure in which the professional performs his work activity and his experiencing feelings of alienation from work. This relationship was explored for each of four empirical indicators of organizational structure for all professionals, and separately for professionals differing in length and type of professional training and for professionals working in two very different organizational units.

In general, the findings of this research support the hypothesis that alienation from work is a consequence of the professional-bureaucratic dilemma for industrial scientists and engineers. Differences in type of supervision, freedom of research choice, professional climate, and company encouragement were associated with degree of work alienation in the expected manner.

The conditioning effects of length and type of professional training on the above relationships were only partially supported by these data. Alienation from work was more strongly associated with type of supervisor and degree of company encouragement among scientists and professionals with advanced training than for engineers and professionals with less advanced training. Freedom of research choice and professional climate were strongly associated with work alienation for all professionals. Moreover, these relationships remained strong when length and type of professional training were controlled.

This unanticipated finding is important when it is recalled that the largest group of professionals employed by the organization are engineers with M.A. or M.S. degrees. These data suggest, therefore, that research freedom and professional atmosphere are more important to the majority of professional personnel than are type of supervision and specific professional incentives (as these factors are related to the experiencing of work alienation).

Although the relationships involving research freedom and professional atmosphere are *similar* for both scientists and engineers, these professionals may be experiencing work alienation for *different* reasons. Freedom of research choice, "having time to pursue one's own research interests," and having "opportunities to obtain professional recognition outside the company" may be interpreted by these professionals in terms of different goals and reference groups. If scientists and engineers differ in their professional goals, then the alienation manifested by engineers may result from their lack of power and participation in organizational affairs whereas the alienation manifested by scientists may reflect their lack of autonomy to pursue their work with the passion Weber described. This interpretation is consistent with the findings obtained in the relationships involving company encouragement and type of supervisor. That these relationships were stronger for the scientists and for those with advanced training may reflect the facts that the specific incentives comprising the company encouragement index are more important for the attainment of scientific than organizational goals, and that working for Participatory supervisors reflects participation in the decision-making process for engineers and violation of professional freedom for scientists who prefer to be left alone as much as possible.

Striking differences were found in the organizational structures of the Aero-Space Group and the Basic Science Laboratory. In addition, degree of work alienation was found to be highly related to type of organizational unit, with Laboratory personnel experiencing a very low degree of alienation as compared with Aero-Space personnel. This relationship remained consistent with respect to direction and degree when length and type of professional training were controlled.

The very different organizational structures evident in the Laboratory and the Aero-Space Group, as well as the effect of such structures upon work alienation for different types of professionals within these units, raise questions concerning the empirical generality of many studies to date which have focused on the Laboratory rather than the larger unit in their analysis. This research suggests that caution should accompany attempts to generalize across units within an organization. Studies of Laboratory personnel may yield findings very different from those obtained from professionals working in the larger organization—where most professional scientists and engineers are in fact employed.

NOTES

1. Amitai Etzioni, *Modern Organizations*, Englewood Cliffs: Prentice-Hall, 1964, p. 1.
2. William Kornhauser, *Scientists in Industry: Conflict and Accommodation*, Berkeley: University of California Press, 1962, p. 10.
3. Simon Marcson, *The Scientist in American Industry*, New York: Harper and Brothers, 1960, p. 5.
4. Arthur L. Stinchcombe, "The Sociology of Organization and the Theory of the Firm," *Pacific Sociological Review*, 13 (Fall, 1960), p. 80.
5. National Science Foundation, *Scientific and Technical Personnel in Industry 1960*, Washington, D.C.: Government Printing Office, 1961, p. 20.
6. National Science Foundation, *Research and Development in Industry 1960*, Washington, D.C.: Government Printing Office, 1963, p. 64.
7. Peter M. Blau and W. Richard Scott, *Formal Organizations*, San Francisco: Chandler Publishing Company, 1962, p. 60.
8. Kornhauser, *op. cit.*, pp. 195–196.
9. These are two of the four areas of conflict identified and discussed by William Kornhauser. See Kornhauser, *op. cit.*, p. 45.
10. Charles D. Orth, "The Optimum Climate for Industrial Research," in Norman Kaplan, ed., *Science and Society*, Chicago: Rand McNally, 1965, p. 141.
11. Kornhauser, *op. cit.*, p. 138.
12. Howard S. Becker and James Carper, "The Elements of Identification with an Occupation," *American Sociological Review*, 21 (June, 1956), pp. 341–348.
13. Clovis Shepherd, "Orientations of Scientists and Engineers," *Pacific Sociological Review*, 4 (Fall, 1961), pp. 79–83.
14. Fred H. Goldner and R. R. Ritti, "Professionalization as Career Immobility," *The American Journal of Sociology*, 72 (March, 1967), pp. 491–494.
15. Harold L. Wilensky, *Intellectuals in Labor Unions: Organizational Pressures on Professional Roles*, Glencoe: The Free Press, 1956.
16. Paula Brown, "Bureaucracy in a Governmental Laboratory," *Social Forces*, 32 (March, 1954), pp. 259–268.
17. Donald Pelz and Frank M. Andrews, "Organizational Atmosphere, Motivation and Research Contribution," *The American Behavioral Scientist*, 6 (December, 1962), pp. 43–47. See also, *The Conflict Between the Scientific and the Management Mind*, Princeton: Opinion Research Corporation, 1959.
18. Etzioni, *op. cit.*, p. 75.
19. Kornhauser, *op. cit.*, pp. 148–149.
20. These objectives were emphasized by the director of the laboratory in conversations with the author and appear in various company documents.
21. Amitai Etzioni, *A Comparative Analysis of Complex Organizations*, Glencoe: The Free Press, 1961, p. 53.
22. L. H. Orzack, "Work as a Central Life Interest of Professionals," *Social Problems*, 7 (1959), pp. 125–132.

23. Hans H. Gerth and C. Wright Mills, eds., *From Max Weber: Essays in Sociology*, New York: Oxford University Press, 1946, p. 135.

24. Melvin Seeman, "On the Meaning of Alienation," *American Sociological Review*, 24 (December, 1959), p. 790.

25. Mills argues that to equate alienation with what is commonly measured as "job dissatisfaction" by psychologists and sociologists is to misunderstand Marx. See C. Wright Mills, *The Marxists*, New York: Dell Publishing Company, 1962, p. 86. Harold Wilensky makes the same point and has developed a measure specifically directed toward ascertaining the social-psychological aspects of alienation from work within a framework of role-self analysis. See Harold L. Wilensky, "Work as a Social Problem," in Howard S. Becker, editor, *Social Problems: A Modern Approach*, New York: John Wiley and Sons, 1966, pp. 138–142. However, there does not appear to be agreement concerning this point. Aiken and Hage, for example, have recently developed a measure of alienation from work based on a factor analysis of items concerned with the "degree of satisfaction with various aspects of the respondents' work situation." See Michael Aiken and Jerald Hage, "Organizational Alienation: A Comparative Analysis," *American Sociological Review*, 31 (August, 1966), p. 501.

26. Nancy Morse, *Satisfactions in the White Collar Job*, Ann Arbor: University of Michigan Press, 1953.

27. The best description of the Goodenough technique is Allen L. Edwards, *Techniques of Attitude Scale Construction*, New York: Appleton-Century-Crofts, 1957, pp. 184 ff. Edwards states that, unlike the C. R. produced by the Cornell technique, the C. R. produced by the Goodenough technique *accurately* represents the extent to which individual responses can be reproduced from scale scores. The Coefficient of Scalability is described in Herbert Menzel, "A New Coefficient for Scalogram Analysis," *Public Opinion Quarterly*, 17 (Summer, 1953), pp. 268–280. The Coefficient of Sharpness is described by James A. Davis, "On Criteria for Scale Relationships," *The American Journal of Sociology*, 63 (January, 1958), pp. 371–380.

28. Response categories were reversed in the first item to make them consistent with the others. Persons disagreeing or strongly disagreeing with each item were coded as "1" and those agreeing or strongly agreeing were coded as "0." Thus, a scale score of 5 means that the person "disagreed" with all five items and a scale score of 0 means that he "agreed" with all five items.

29. Etzioni, *Modern Organizations, op. cit.*, p. 80.

30. See for example, Eliot Freidson and Buford Rhea, "Knowledge and Judgment in Professional Evaluations," *Administrative Science Quarterly*, 10 (June, 1965), pp. 107–108, Ernest Greenwood, "Attributes of a Profession," *Social Work*, 2 (July, 1957), pp. 45–55, and Everett C. Hughes, *Men and Their Work*, Glencoe: The Free Press, 1958.

31. Howard Baumgartel, "Leadership, Motivations, and Attitudes in Research Laboratories," *Journal of Social Issues*, 12 (1956), p. 30, and "Leadership Style as a Variable in Research Administration," *Administrative Science Quarterly*, 2 (December, 1957), pp. 344–360.

32. Chris Argyris, *Understanding Organizational Behavior*, Homewood: The Dorsey Press, 1960, p. 94. Italics added. Aiken and Hage found that rule observation, which implies close supervision by superiors, was the single best predictor of alienation from expressive relations. See Aiken and Hage, *op. cit.*, p. 506.

33. Leo Meltzer, "Scientific Productivity in Organizational Settings," *Journal of Social Issues*, 12 (1956), pp. 32–40.

34. See, for example, John W. Riegal, *Intangible Rewards for Engineers and Scientists*, Ann Arbor: University of Michigan Press, 1958, pp. 12–13, and Todd LaPorte,

"Conditions of Strain and Accommodation in Industrial Research Organizations," *Administrative Science Quarterly*, 10 (June, 1965), pp. 33–34.

35. Kornhauser, *op. cit.*, p. 135. The dependence of staff on line is discussed in Melville Dalton, *Men Who Manage*, New York: John Wiley and Sons, 1959.

36. This type of analysis controls for the compositional effect in terms of the similarity in length and type of professional training of the professionals employed by each unit. It should be indicated that differences in recruitment procedures, in professional ability, and in type of work actually engaged in by the professionals may also yield differences in the kinds of professionals employed by each unit and thus affect the relationship between their perceptions of the organizational structure and degree of work alienation.

11

MICHAEL AIKEN AND JERALD HAGE

ORGANIZATIONAL INTERDEPENDENCE AND INTRA-ORGANIZATIONAL STRUCTURE*

The major purpose of this paper is to explore some of the causes and consequences of organizational interpendence among health and welfare organizations. The aspect of organizational interpendence that is examined here is the joint cooperative program with other organizations. In particular, we are interested in relating this aspect of the organization's relationships with its environment to internal organizational behavior.

Thus this paper explores one aspect of the general field of interorganizational analysis. The effect of the environment on organizational behavior as well as the nature of the interorganizational relationships in an organization's environment are topics that have received increasing attention from scholars in recent years. Among studies in the latter category, there are those that have attempted to describe the nature of organizational environments in terms of the degree of turbulence (Emery and Trist 1965; cf. Terreberry, 1968) and in terms of organizational sets (Evan 1966). Others have emphasized transactional interdependencies among orga

Reprinted from Michael Aiken and Jerald Hage, "Organizational Interdependence and Intra-organizational Structure," *American Sociological Review*, 33–6 (December, 1968) pp. 912–929.

* This is a revised version of a paper read at the annual meetings of the American Sociological Association, San Francisco, California, August 30, 1967. This investigation was supported in part by a research grant from the Vocational Rehabilitation Administration, Department of Health, Education, and Welfare, Washington, D.C. We are grateful to Charles Perrow for helpful comments on an earlier version of this paper. In addition, we would like to acknowledge the cooperation and support of Harry Sharp and the Wisconsin Survey Laboratory during the interviewing phase of this project.

izations (Selznick, 1949; Ridgeway, 1957; Dill, 1962; Levine and White, 961; Levine et al., 1963; Guetzkow, 1966; Litwak and Hylton, 1962; James Thompson, 1962; Elling and Halbsky, 1961; Reid, 1964). Still others have mphasized the importance of an understanding of interorganizational elationships for such problem areas as education (Clark, 1965), medical are (Levine and White, 1963), rehabilitation and mental health (Black nd Kase, 1963), delinquency prevention and control (Miller, 1958; Reid, 964); services for the elderly (Morris and Randall, 1965); community ction (Warren, 1967); and community response to disasters (Form and Nosow, 1958).

Few studies, however, have examined the impact of the environment n internal organizational processes. One such study by Thompson and McEwen (1958) showed how the organizational environment can affect goal-setting in organizations, while a study by Dill (1958) examined how nvironmental pressures affect the degree of managerial autonomy. Simpon and Gulley (1962) found that voluntary organizations with diffuse ressures from the environment were more likely to have decentralized tructures, high internal communications, and high membership involvenent, while those having more restricted pressures from the environment ad the opposite characteristics. Terreberry (1968) has hypothesized that rganizational change is largely induced by forces in the environment, nd Yuchtman and Seashore (1967) have defined organizational effectiveess in terms of the organization's success in obtaining resources from he environment. Recently, James D. Thompson (1967) and Lawrence nd Lorsch (1967) have suggested some ways in which elements in the nvironment can affect organizational behavior. There are also other tudies which argue that another aspect of the environment—variations in ultural values and norms—may also affect the internal structure of orgaizations (Richardson, 1959; Harbison et al., 1963; Crozier, 1964). Each f these studies, then, suggests ways in which the organization's environnent affects the internal nature of the organization. The purpose of this tudy is to show how one aspect of the organization's relationship with its nvironment, i.e., the interdependence that arises through joint cooperaive programs with other organizations, is related to several intra-organizaional characteristics. We shall do this by describing a theoretical ramework about organizational interdependence and then by examining ome results from an empirical study of organizational interdependence.

A second objective in calling attention to this relatively neglected area f organizational analysis is to suggest that the processes of both conflict nd cooperation can be incorporated into the same model of organizational interdependence. The concept of interdependence helps us to focu on the problem of interorganizational exchanges. At the same time, th exchange of resources, another aspect of the relationships between organ zations, is likely to involve an element of conflict. While Simmel has mad the dialectic of cooperation and conflict a truism, as yet there has bee little work that explains interorganizational cooperation and conflict. Cap low (1964) has suggested a model of conflict involving the variables o subjugation, insulation, violence, and attrition, but this model focuse neither on the particular internal conditions that give rise to intero ganizational relationships nor on the consequences of them for organiza tional structure. These are key intellectual problems in attempting t understand exchanges among organizations.

The models of pluralistic societies described by Tocqueville (1945 and more recently by Kornhauser (1959) underscore the importance c autonomous and competing organizations for viable democratic processe Such theoretical models assume that the processes of conflict as well a cooperation inhere in social reality. Recent American social theory ha been criticized for its excessive emphasis on a static view of socia processes and for failing to include conflict in its conceptual model (Dahrendorf, 1958; Coser, 1956; Wrong, 1961). The study of interorgan zational relationships appears to be one area which can appropriatel incorporate the processes of both conflict and cooperation. Therefore th concept of organizational interdependence becomes a critical analytica tool for understanding this process.

Most studies of organizational interdependence essentially conceive c the organization as an entity that needs inputs and provides output linking together a number of organizations via the mechanisms c exchanges or transactions. (Cf. Ridgeway, 1957; Elling and Halbsky, 196 Levine and White, 1961; Dill, 1962; James D. Thompson, 1962.) Som types of organizational exchanges involve the sharing of clients, fund and staff in order to perform activities for some common objective (Levin et al., 1963). The measure of the degree of organizational interdependenc used here is the *number of joint programs* that a focal organization ha with other organizations. The greater the number of joint programs, th more organizational decision-making is constrained through obligation commitments, or contracts with other organizations, and the greater th degree of organizational interdependence. (Cf. Guetzkow, 1966.) Th type of interdependence among health and welfare organizations ha variously been called "functional co-operation" by Black and Kas (1963), and "program co-ordination" by Reid (1964), and is considere more binding form of interdependence and therefore a more interesting xample of interorganizational cooperation. This does not suggest that the ooperation that is involved in joint programs is easily achieved. On the ontrary, there are a number of barriers to establishing such interde- endencies among organizations (cf. Johns and de Marche, 1951), and he probability of conflict is quite high, as Miller (1958) and Barth 1963) point out.

The reader may wonder why the concept of the joint program is appar- ntly such an important kind of interorganizational relationship. The nswer is that, unlike exchanges of clients or funds (which may only nply the *purchase* of services) or other types of organizational coopera- on, a joint program is often a relatively enduring relationship, thus ndicating a high degree of organizational interdependence.

The *joint program* needs to be carefully distinguished from the *joint rganization*. The latter refers to the situation in which two or more rganizations create a separate organization for some common purpose. 'or example, the Community Chest has been created by health and wel- are organizations for fund-raising purposes. Similarly, Harrison (1959) as noted that the Baptist Convention was created by the separate Baptist hurches for more effective fund raising. Guetzkow (1950) has described nteragency committees among federal agencies, representing a special ase of the joint organization. Business firms have created joint organiza- ions in order to provide service functions. These are clearly different from he joint program because these joint organizations have separate corporate lentities and often their own staff, budget, and objectives.

Some examples of joint programs in organizations other than those in he health and welfare field are the student exchange programs in the ig Ten. Harvard, Columbia, Yale, and Cornell Universities are develop- ng a common computerized medical library. Indeed, it is interesting to ote how many universities use joint programs of one kind or another. We o not believe that this is an accident; rather, it flows from the character- tics of these organizations. In our study, which includes rehabilitation enters, we have observed the attempt by one organization to develop a umber of joint programs for the mentally retarded. These efforts are eing financed by the Department of Health, Education, and Welfare, nd evidently reflect a governmental concern for creating more coopera- ive relationships among organizations. Even in the business world, where he pursuit of profit would seem to make the joint program an impossibil- y, there are examples of this phenomenon. Recently, Ford and Mobil)il started a joint research project designed to develop a superior gaso-

line. This pattern is developing even across national boundaries in bot
the business and nonbusiness sectors.

It is this apparently increasing frequency of joint programs that make
this form of interdependence not only empirically relevant, but theoreti
cally strategic. In so far as we can determine, organizational interde
pendence is increasingly more common (Terreberry, 1968), but the ques
tion of why remains to be answered.

Theoretical Framework

The basic assumptions that are made about organizational behavior an
the hypotheses of this study are shown in Table 11–1. These assumption
provide the argument, or model, to use Willer's (1967) term, for th
hypotheses to be tested below.

The first three assumptions deal with the basic problem of why organiza
tions, at least health and welfare organizations, become involved in inte
dependent relationships with other units. The type of interdependenc
with which we are concerned here is the establishment of joint, coopera
tive activities with other organizations. If we accept Gouldner's (1959
premise that there is a strain toward organizations maximizing thei
autonomy, then the establishment of an interdependency with anothe
organization would seem to be an undesirable course of action. It is th

TABLE 11-1
Assumptions and Hypotheses about Organizational Interdependence

Assumptions:

I. Internal organizational diversity stimulates organizational innovation.
II. Organizational innovation increases the need for resources.
III. As the need for resources intensifies, organizations are more likely to develop greater interdependencies with other organizations, joint programs, in order to gain resources.
IV. Organizations attempt to maximize gains and minimize losses in attempting to obtain resources.
V. Heightened interdependence increases problems of internal control and coordination.
VI. Heightened interdependence increases the internal diversity of the organization.

Hypotheses:

1. A high degree of complexity varies directly with a high number of joint programs.
2. A high degree of program innovation varies directly with a number of joint programs.
3. A high rate of internal communication varies directly with a high number of joint programs.
4. A high degree of centralization varies inversely with a high number of joint programs.
5. A high degree of formalization varies inversely with a high number of joint programs.

view here that organizations are "pushed" into such interdependencies because of their need for resources—not only money, but also resources such as specialized skills, access to particular kinds of markets, and the like (cf. Levine et al., 1963).

One source of the need for additional resources results from a heightened rate of innovation, which in turn is a function of internal organizational diversity. In several ways internal diversity creates a strain towards innovation and change. The conflict between different occupations and interest groups, or even different theoretical, philosophical, or other perspectives, results in new ways of looking at organizational problems. The likely result of this is a high rate of both proposals for program innovations as well as successful implementation of them (Hage and Aiken, 1967). But organizational diversity also implies a greater knowledge and awareness of the nature of and changes in the organizational environment, particularly when organizational diversity implies not only a spectrum of occupational roles in the organization, but also involvement in professional societies in the environment by the incumbents of those occupational roles, itself a type of organizational interdependency. Together the internal conflicts and awareness of the nature of the organization's environment create strains towards organizational change.

But innovation has its price. There is a need for more resources to pay the costs of implementing such innovations—not only money, but staff, space, and time. The greater the magnitude of the change or the number of changes within some specified period of time, the greater the amounts of resource that will be needed and the less likely that the normal sources will be sufficient. Some have called organizations that successfully accomplish this task effective ones (Yuchtman and Seashore, 1967). Thus, the leaders of innovating organizations must search for other possibilities, and the creation of a joint, cooperative project with another organization becomes one solution to this problem.

This mechanism for gaining resources, i.e., the establishment of a joint program, is best viewed as a type of organizational exchange. The leaders sacrifice a small amount of autonomy for gains in staff, funds, etc. While there are strong organizational imperatives against such exchanges, since they inevitably involve some loss of autonomy, as well as necessitate greater internal coordination, the increased intensification of needs for greater resources makes such an alternative increasingly attractive. Still another factor involved here is that some objectives can only be achieved through cooperation in some joint program. The goal may be so complicated or the distribution of risk so great that organizations are impelled to enter into some type of joint venture. Of course the creation of inte dependencies with other organizations also has its costs. The organizatio must utilize some of its own resources in order to perform whateve coordination is necessary. Hence an organization with no surplus resource available could hardly afford a joint program. Thus there must be som slack in the resource base in the organization before any innovation o cooperative venture is likely.

This is not to argue for the perfect rationality of organizational leaders Some decisions about change or the choice of a cooperative activity ma be quite irrational, and perhaps non-logical (Wilensky, 1967). Indee much of our argument about the conditions that lead to organizationa innovation, i.e., conflict among different occupations, interest groups, o perspectives, is that this is hardly the most rational way to bring abou change. Perhaps it is best to view the process as a series of circumstance that propel such events.

While we feel that this line of reasoning is a valid explanation of wh organizations enter into interdependent relationships with other organiza tions via such mechanisms as the joint program, alternative explanation have been offered and must be considered. Lefton and Rosengren (1966 have suggested that the lateral and longitudinal dimensions of organiza tional commitment to clients are factors, at least in health and welfar organizations. These are probably not the primary factors in other type of organizations, such as economic ones. However, our concern has bee to attempt to find the most general argument possible to explain organi zational interdependence. At the same time we have left unanswered th question of why organizations become diverse in the first place, and thei framework may provide one possible answer. Reid (1964) has indicate that complementary resources are also an important factor in understand ing organizational interdependence. Without necessarily agreeing or dis agreeing with these points of view, we do believe that the first thre assumptions in Table 11–1 represent *one* causal chain showing wh organizations become involved in more enduring interorganizational rela tionships.

The next theoretical problem is what kind of organization is likely t be chosen as a partner in an interdependent relationship. Here we assum that organizations attempt to maximize their gains and minimize thei losses. This is our fourth premise. That is, they want to lose as littl power and autonomy as possible in their exchange for other resource This suggests that they are most likely to choose organizations with com plementary resources, as Reid (1967) has suggested, or partners wit different goals, as Guetzkow (1966) has indicated. This reduces some of the problem of decreased autonomy, because the probability of conflict is reduced and cooperation facilitated in such symbiotic arrangements (cf. Hawley, 1951). This assumption also implies that other kinds of strategies might be used by the leaders of the organization once they have chosen the joint program as a mechanism of obtaining resources. Perhaps it is best to develop interdependent relationships with a number of organizations in order to obtain a given set of resources, thus reducing the degree of dependence on a given source. Again, we do not want to argue that organizational leaders will always choose the rational or logical alternative, but rather that they will simply *attempt* to minimize losses and maximize gains. Under circumstances of imperfect knowledge, some decisions will undoubtedly be irrational.

Our last theoretical problem is consideration of the consequences for the organization of establishing interdependent relationships as a means of gaining additional resources. Such joint activities will necessitate a set of arrangements between the participating organizations to carry out the program. This will mean commitments to the other organization, resulting in constraints on some aspects of organizational behavior. This in turn will mean an increase in problems of internal coordination, our fifth assumption. It is often difficult to work with outsiders, i.e., the partner in a joint activity. In this circumstance a number of mutual adaptations in a number of different areas will become necessary. One solution to this problem is the creation of extensive internal communication channels, such as a broad committee structure which meets frequently.

But perhaps a more interesting consequence of the joint program is that it can in turn contribute to organizational diversity. There is not only the likelihood of the addition of new staff from other organizations, but, more importantly, the creation of new communication links with other units in the organization's environment. New windows will have been opened into the organization, infusing new ideas and feeding the diversity of the organization, which means that the cycle of change, with all of its consequences, is likely to be regenerated.

In this way a never-ending cycle of diversity—innovation—need for resources—establishment of joint programs—is created. What may start as an interim solution to a problem can become a long-term organizational commitment which has a profound impact on the organization. In the long run, there is the tendency for units in an organizational set to become netted together in a web of interdependencies. (Cf. Terreberry, 1968.)

With these six assumptions, a large number of testable hypotheses can be deduced. Indeed this is one of the advantages of a general theoretical framework. Not only does it provide the rationale for the hypotheses being tested, but it can suggest additional ideas for future research. Since we are mainly concerned with the factors associated with high interdependency, and more particularly the number of joint programs, all of the hypotheses in Table 11–1 are stated in terms of this variable.

Organizational diversity implies many different kinds of variables. We have examined three separate indicators of it: diversity in the number of occupations or the degree of complexity; diversity in the number of power groups or the degree of centralization; and diversity in the actual work experience or the degree of formalization. If Assumptions I–III are correct, then the stimulation of change, and more particularly innovation brought about by each of these kinds of diversity, should be associated with a large number of programs. But this is not the only way in which these variables can be related; and that observation only emphasizes how the internal structure of the organization affects the extent of the enduring relationships with other organizations. The problems of internal coordination and the increased diversity, Assumptions V and VI, are also related. Both mechanisms of coordination—communication and programming—are undoubtedly tried, but communication is probably preferred. This increases the advantages of diversity and also helps to bring about greater decentralization and less formalization. Similarly, the greater awareness of the environment, via the infusion of staff from other organizations, feeds this cycle of cause and effect relationships. Therefore, we have hypothesized that the number of joint programs varies directly with the degree of complexity (Hypothesis 1) and inversely with the degree of centralization and formalization (Hypotheses 4 and 5).

Since our arguments also involve statements about the stimulation of innovation, which in turn heightens the need for resources, it is clear that we would expect the degree of innovation to co-vary with the number of joint programs. This is Hypothesis 2 of Table 11–1. While program change is only one kind of organizational innovation, it is probably the most important, at least from the standpoint of generating needs for additional resources, and thus it goes to the heart of the argument presented in Table 11–1. Program innovation in turn has consequences for the degree of centralization and formalization in the organization, but here we are mainly concerned about the relationship between the rate of organization innovation as reflected in new programs and the number of joint programs, and not about these other mediating influences.

The degree of attempted internal coordination is measured by only one variable, namely the rate of communication, but again we feel that this is an important indication of this idea. Given the desire to minimize the loss of autonomy (Assumption IV), organizational members must be particularly circumspect when dealing with staff and other kinds of resources from their organizational partners. This largely reduces the options about programming and encourages the elite to emphasize communication rates. Probably special "boundary spanning" roles (Thompson, 1962) are created; these men negotiate the transactions with other organizations and in turn keep their organizational members informed. The problems of interpenetration by other organizational members will keep the communication channels open and filled with messages as internal adjustments are made. Thus this is the rationale for the third hypothesis.

Study Design and Methodology

The data upon which this study is based were gathered in sixteen social welfare and health organizations located in a large midwestern metropolis in 1967. The study is a replication of an earlier study conducted in 1964. Ten organizations were private; six were either public or branches of public agencies. These organizations were all the larger welfare organizations that provide rehabilitation, psychiatric services, and services for the mentally retarded, as defined by the directory of the Community Chest. The organizations vary in size from twenty-four to several hundred. Interviews were conducted with 520 staff members of these sixteen organizations. Respondents within each organization were selected by the following criteria: (1) all executive directors and department heads; (2) in departments of less than ten members, one-half of the staff was selected randomly; (3) in departments of more than ten members, one-third of the staff was selected randomly. Non-supervisory administrative and maintenance personnel were not interviewed.

AGGREGATION OF DATA

This sampling procedure divides the organization into levels and departments. Job occupants in the upper levels were selected because they are most likely to be key decision-makers and to determine organizational policy, whereas job occupants on the lower levels were selected randomly. The different ratios within departments ensured that smaller departments were adequately represented. Professionals, such as psychiatrists, social workers and rehabilitation counselors, are included because they are intimately involved in the achievement of organizational goals and are likely to have organizational power. Non-professionals, such as attendants, janitors, and secretaries are excluded because they are less directly involved in the achievement of organizational objectives and have little or no power. The number of interviews varied from eleven in the smallest organization to sixty-two in one of the larger organizations.

It should be stressed that in this study the units of analysis are *organizations*, not individuals in the organizations. Information obtained from respondents was pooled to reflect properties of the sixteen organizations, and these properties were then related to one another. Aggregating individual data in this way presents methodological problems for which there are yet no satisfactory solutions. For example, if all respondents are equally weighted, undue weight is given to respondents lower in the hierarchy. Yet those higher in the chain of command, not the lower-status staff members, are the ones most likely to make the decisions which give an agency in ethos.[1]

We attempted to compensate for this by computing an organizational score from the means of social position within the agency. A social position is defined by the level or stratum in the organization and the department or type of professional activity. For example, if an agency's professional staff consists of psychiatrists and social workers, each divided into two hierarchical levels, the agency has four social positions: supervisory psychiatrists, psychiatrists, supervisory social workers, and social workers. A mean was then computed for each social position in the agency. The organizational score for a given variable was determined by computing the average of all social position means in the agency.[2]

The procedure for computing organizational scores parallels the method utilized in selecting respondents. It attempts to represent organizational life more accurately by not giving disproportionate weight to those social positions that have little power and that are little involved in the achievement of organizational goals.

Computation of means for each social position has the advantage of avoiding the potential problem created by the use of different sampling ratios. In effect, responses are standardized by organizational location—level and department—and then combined into an organizational score. Computation of means of social position also has a major theoretical advantage in that it focuses on the sociological perspective of organizational reality.

We make no assumption that the distribution of power, regulations, or rewards is random within any particular social position. Instead, each respondent is treated as if he provides a true estimate of the score for a given social position. There is likely to be some distortion due to personality differences or events unique in the history of the organization, but the computation of means for each social position hopefully eliminates or at least reduces the variation due to such factors. By obtaining measures from all levels and all departments, the total structure is portrayed and reflected in the organizational score.

THE MEASUREMENT OF ORGANIZATIONAL INTERDEPENDENCE

The degree of organizational interdependence is measured by the number of joint programs with other organizations. There are several possible measures of the nature and degree of organizational interdependence among social welfare and health organizations. Among these are:

1. The number of cases, clients or patients referred or exchanged.
2. The number of personnel lent, borrowed, or exchanged.
3. The number, sources, and amounts of financial support.
4. The number of joint programs.

The first two of these were used in an earlier study of interorganizational relationships (Levine and White, 1961). In our research we found that organizations such as rehabilitation workshops and family agencies simply did not keep records of the number of walk-ins or calls referred by other organizations. Similar problems were encountered with exchanges of personnel. Thus, we found great difficulty in using these measures of interdependence. While the nature and amounts of financial support are interesting and important aspects of interorganizational analysis, they are not included in this study.

We asked the head of each organization to list every joint program in which his organization had been involved in the past ten years, whether terminated or not. A profile of each program was obtained, including the names of participating organizations, goals of the program, number and type of clients or patients involved, and source of financial and other resources for the program. Only existing programs and those involving the commitment of resources by all participating organizations—such as personnel, finances, space—were included in our analysis.

Since a number of our sixteen organizations had participated in joint programs with each other, it was possible to check the reliability of their responses. We did not find any difficulties of recall for this period of time. In part this is probably because most of the joint programs, once started, tended to continue over time. Some organizations had maintained their organizational relationships for as many as twenty years. Then too, the fact that the joint program is not a minor incident in the life of an organization also facilitates recall. We did discover that organizational leaders tended to think of the purchase of services as a joint program. To solve this problem we included in our interview schedule a series of follow-up questions about the amount of staff shared and the amount of funds contributed by each organization involved in the joint program.

Another problem of measurement centered on the difficulty of defining separate joint programs. For example, there was a tendency for an organization with a history of successful relationships (those that endured for more than two years) to develop a number of joint programs with the same organization. The relationships would grow in scope and depth in much the way that one would predict from Homans' (1950) hypotheses about the interaction between people. This raised the problem of whether joint programs with the same organization should be counted as separate programs. Our solution was to count the program separately if it involved different activities. Thus a research program and an education program with the same organization, two common kinds of programs, would be counted as separate joint programs. The key in making this decision was the idea of separate activities. In fact, programs were usually developed at different dates, suggesting again that our solution was a correct one. At the same time, if an organization developed the same joint program with three organizations, this was counted only once. From a practical standpoint these attempts at refinement were not so important because it is

TABLE 11-2
Average Number of Joint Programs by Type of Organization

Type of organizations	*Number of organizations*	*Average number of joint programs*	*Range*
Rehabilitation centers	3	20.7	8-33
Special education department—public schools	1	15.0	15
Hospitals	3	8.3	6-12
Homes for emotionally disturbed	3	2.3	1-3
Social casework agencies	6	1.2	0-4
All organizations	16	7.3	0-33

clear that the differences in number of joint programs among the sixteen organizations in our study are so great that similar ranking would occur regardless of how one counted the programs.

The number of existing joint programs among these sixteen organizations ranged from none to 33. Rehabilitation centers had the highest average number of joint programs, although the range was quite extensive among some other kinds of organizations in our study (Table 11–2). The special education department and the hospitals had an intermediate range of programs. Social casework agencies and homes for the emotionally disturbed had the least number of joint programs. In every case, however, there was some variation within each organizational category.

Findings

A strict interpretation of data would allow us to discuss only the consequences of interorganizational relationships on the internal structure and performance of an organization. This is true because the period of time during which measurement of the number of joint programs, our measure of organizational interdependence, was made occurred prior to most of our measures of structure and performance. Yet the reasoning in our theoretical framework suggests that these variables are both causes and effects in an on-going process. Strictly speaking, our data reflect the consequences of increased joint programs, but we shall still make some inferences about their causes.

1. *Organizations with many joint programs are more complex organizations, that is, they are more highly professionalized and have more diversified occupational structures.* By complexity we do not mean the same thing as Rushing's (1967) division of labor, a measure of the distribution of people among different occupations, but rather the diversity of activities. There are essentially two aspects of complexity as we have defined it: the degree to which there is a high number of different types of occupational activities in the organization; and the degree to which these diverse occupations are anchored in professional societies.[3] One of the most startling findings in our study is the extremely high correlation between the number of different types of occupations in an organization and the number of joint programs ($r = 0.87$).

The relationship between the occupational diversity of the organization and the number of joint programs in 1967 is very high, whether we use the number of occupations in 1959 ($r = 0.79$), the number of occupations in 1964 ($r = 0.83$), or the number of occupations in 1967 ($r = 0.87$). While time sequence is not the same as causation, this does suggest that occupational diversity is not solely a function of new programs. Rather it suggests that organizations that have a high number of joint programs are organizations that have been occupationally diverse for a number of years.

The addition of joint programs evidently makes an organization aware of the need for still more specialities. One rehabilitation center used social workers in a joint program involving the mentally retarded with several other agencies. It then decided to add social workers to a number of its other programs. The addition of new specialties may also be neces-

TABLE 11-3
Relationships between the Number of Joint Programs and Organizational Characteristics

Organizational characteristics	*Pearsonian product-moment correlation coefficients between each organizational characteristic and the number of joint programs*
1. *Degree of complexity*	
Index of professional training	.15
Index of professional activity	.60**
Number of occupations: 1967	.87****
2. *Degree of organizational innovation: 1959-1966*	
Number of new programs (including new programs that are joint programs)	.71***
Number of new programs (excluding new programs that are joint programs)	.74****
3. *Internal communication*	
Number of committees	.47*
Number of committee meetings per month	.83****
4. *Degree of centralization*	
Index of participation in decision-making	.30
Index of hierarchy of authority	.33
5. *Degree of formalization*	
Index of job codification	.13
Index of rule observation	−.06
Index of specificity of job	−.06

*$P < .10$.
**$P < .05$.
***$P < .01$.
****$P < .001$.

TABLE 11-4
Comparison of Pearsonian Correlation Coefficient (r), Spearman's Rank Order Correlation Coefficient (rho), and Kendall's Rank Correlation Coefficient (tau) for the Four Largest Correlations Shown in Table 11-3

	Correlation coefficient between number of joint programs and organizational characteristics		
Organizational characteristics	*r*	*rho*	*tau*
Number of occupations: 1967	.87	.81	.74
Number of new programs: 1959-1966 (including new programs that are joint programs)	.71	.84	.75
Number of new programs: 1959-1966 (excluding new programs that are joint programs)	.74	.80	.70
Number of committee meetings per month	.83	.61	.54

ary in order to help solve some of the problems of coordination created by
ie joint programs.

The dependent variable, number of joint programs, is quite dispersed
ith a range from 0 to 33 and a mean of 7.3. It is entirely possible that
ie unusually high correlations for some variables in Table 11–3 are sim-
ly a function of a highly skewed distribution on this variable. Therefore,
e computed two non-parametric measures of correlation, Spearman's
ank order correlation coefficient (rho) and Kendall's rank correlation
oefficient (tau) for the relationship between number of occupations in
967 and the number of joint programs as shown in Table 11 4. The
elationship between these two variables remains strong even when using
ie non-parametric statistics.

The objection could be raised that the very strong relationship between
umber of occupational specialties and the number of joint programs may
lso be a function of the type of organization. In Table 11–2, it was
iown that rehabilitation centers had the most joint programs, followed by
ie special education department, hospitals, homes for the emotionally
isturbed, and finally social casework agencies. The observation that there
a positive relationship between these two variables is valid within
iree of the four categories of organizations shown in Table 11–5.

That is, within the categories of rehabilitation centers, mental hospitals,
id homes for the emotionally disturbed the organizations having the
ighest number of occupations have the most joint programs while those
aving the fewest occupational specialties have the smallest number of

TABLE 11-5
Number of Occupations in 1967 and Number of Joint Programs by Type of Organization

	Number of occupations 1967	*Number of joint programs*
Rehabilitation centers		
Rehabilitation center A	27	33
Rehabilitation center B	24	21
Rehabilitation center C	13	8
Department of special education		
Educational organization D	19	15
Mental hospitals		
Mental hospital E	18	12
Mental hospital F	18	7
Mental hospital G	11	6
Homes for emotionally disturbed		
Home H	11	3
Home I	10	3
Home J	7	1
Social casework agencies		
Casework agency K	7	1
Casework agency L	6	0
Casework agency M	5	1
Casework agency N	5	1
Casework agency O	4	4
Casework agency P	1	0

joint programs. Only among social casework agencies does the relationship not hold. It might be noted that only one social casework organization had more than one interorganizational tie.

The degree to which an organization is professionalized is also strongly related to the number of joint programs. We measured the degree of professionalism in organizations in two ways: first, the degree to which the organizational members received professional training; and second, the degree to which organizational members are currently active in professional activities, i.e., attending meetings giving papers, or holding offices. The measure of current professional activity was also quite highly related to our measure of the number of joint programs ($r = 0.60$).[4] The degree of professional training had little relationship with the number of joint programs ($r = 0.15$).[5]

2. *Organizations with many joint programs are more innovative organizations.* The degree of organizational innovation is measured by the number of new programs that were successfully implemented in the organization during the eight-year period from 1959 to 1966. The correlation

coefficient between joint programs and new programs is 0.71, as shown in Table 11–3. Of course, there is an element of spuriousness in this relationship, since some of the new programs were joint programs. If the correlation coefficient is recomputed, eliminating all new programs that are also joint programs, we find the same result ($r = 0.74$).

As in the case of number of occupational specialties in the organization, the finding based on non-parametric measures of association between each of these two measures of organizational innovation and the number of new programs is little different from the results based on the parametric statistical measure (See Table 11–4).

It could be that the above relationships between degree of organizational innovation and number of joint programs may simply be a function of complexity. We have argued that the degree of complexity gives rise not only to joint programs, but also to new programs. While there is no relationship between professional training and the number of new programs ($r = -0.18$), there are relatively strong relationships between this variable and professional activity ($r = 0.74$) as well as occupational diversity ($r = 0.67$). When the relationships between the number of joint programs and the number of new programs (excluding new programs that are joint programs) is controlled for each of these three indicators separately, the relationship between these two variables remains relatively strong (see Table 11–6). This illustrates that the number of new programs is related to the number of joint programs independently of these various indicators of complexity.

The key idea in our interpretation is that it is the rate of organizational

TABLE 11-6
Partial Correlation Coefficients between Number of Joint Programs and Organizational Innovation, Controlling for Indicators of Complexity

Control variables	*Partial correlation between number of joint programs and number of new programs 1959-1966 (excluding new programs that are joint programs), controlling for the variable indicated*
Indicators of complexity	
Index of professional training	.77
Index of professional activity	.55
Number of occupations: 1967	.46

innovation that intensifies the need for new resources. The higher thi rate, the more likely organizations are to use the joint program as mechanism for cost reduction in such activities. The fact that some new programs are joint programs only strengthens our argument that the join program is a useful solution for the organization seeking to develop new programs.

This interplay between new programs and joint programs can be made clear with several examples from our study. One rehabilitation center wit a high rate of new programs developed joint programs with several organi zations that were primarily fund-raising organizations, as a solution fo funding its growth. But in turn these organizations recognized new need and asked the organization to develop still more new programs in area for their clients. This particular agency is presently exploring the possibil ity of developing special toys for the mentally retarded because one o its joint programs is with an organization concerned with this type o client.

We may also re-examine the relationships between indicators of com plexity and the number of joint programs. As shown in Table 11–7, only the relationship between the number of occupations and the number o joint programs remains strong when the number of new programs (ex cluding new programs that are joint programs) is controlled (partia $r = 0.75$).

3. *Organizations with many joint programs have more active interna communication channels.* We measured the degree of internal communi cation in two ways. First, the number of committees in the organizatio

TABLE 11-7
Partial Correlation Coefficients between Number of Joint Programs and Indicators of Complexity, Controlling for Number of New Programs (Excluding New Programs that are Joint Programs)

Indicators of complexity	*Partial correlation between number of joint programs and indicators of complexity, controlling for number of new programs (excluding new programs that are joint programs)*
Index of professional training	.32
Index of professional activity	.11
Number of occupations: 1967	.75

and, second, the number of committee meetings per month. An active committee structure in an organization provides the potential for viable communication links in an organization. As shown in Table 11–3, there was a moderately strong relationship between the number of organizational committees and joint programs ($r = 0.47$) and a very strong relationship between the number of committee meetings per month and the number of joint programs ($r = 0.83$).

The relationship between the number of joint programs and the number of committee meetings per month remains moderately strong when the two non-parametric measures of association are computed. (See Table 11–4.)

Actually the system of communication for joint programs is even more complex than this. For example, one rehabilitation agency with the largest number of joint programs had a special board with the university with which it had many joint programs and was in the process of establishing another joint board with a second university. Another rehabilitation agency created a special steering committee to suggest and supervise joint programs: the members of this committee were representatives from other organizations.

Controlling for the indicators of complexity and program change reduces the relationship between the number of committees and number of joint programs almost to zero in every case except that of professional training. Thus, the number of committees is evidently a function of these factors. On the other hand, the very strong relationship between the number of joint programs and the frequency of committee meetings is only moderately reduced when these controls are applied as shown in Table 11–8. This shows that the frequency of committee meetings is not simply a function of the complexity of the organization or the degree of organizational innovation, but has an independent relationship with the number of joint programs.

4. *Organizations with many joint programs have slightly more decentralized decision-making structures*. In our study, staff members were asked how often they participated in organizational decisions about the hiring of personnel, the promotion of personnel, the adoption of new organizational policies, and the adoption of new programs or services. The organizational score was based on the degree of participation in these four areas of decision-making.[6] As shown in Table 11–3, there is a weak, positive relationship between the degree of participation in agency-wide decisions and the number of joint programs ($r = 0.30$). This appears to be measuring the way resources are controlled. A second kind of decision-

TABLE 11-8
Partial Correlation Coefficients between Number of Joint Programs and Indicators of Internal Communication, Controlling for Indicators of Complexity and Innovation

Control variables	*Partial correlation between number of joint programs and number of committees, controlling for the variable indicated*	*Partial correlation between number of joint programs and frequency of committee meetings, controlling for the variable indicated*
Indicators of complexity		
Index of professional training	.45	.83
Index of professional activity	.13	.76
Number of occupations: 1967	.11	.57
Indicator of organizational innovation		
Number of new programs: 1959-1966 (excluding new programs that are joint programs)	.08	.64

making concerns the control of work. We measure the degree of decision-making about work with a scale called the "hierarchy of authority."[7] This scale had a relationship with the number of joint programs in the opposite direction to our expection (r = 0.33). While highly interdependent organizations have slightly more decentralization of decisions about organizational resources, there is slightly less control over work in such organizations. It is difficult to account for this other than that the organizations with a high degree of program change during the period 1964–1966 had less control over work decisions in 1967 than in 1964. This suggests that the rate of change was so high in such organizations during this period that some more rigid mechanisms of social control were adopted in these organizations. Since the highly innovative organizations were also those with more joint programs, this helps to explain the reversal.

Partial correlations between the number of joint programs and the degree of participation in decision-making, controlling for each of the indicators of complexity, innovation, and internal communication, are shown in Table 11–9.

The relatively low relationship between these two variables is reduced, and in one case reversed, when these other factors are controlled by using partial correlations. Only in the case of frequency of committee meetings is the relationship strengthened. What this means is that the degree of participation in decision-making is largely a function of some of the previ ously discussed variables—professional activity, number of occupations, and number of committees. Thus, it has little independent relationship with the number of joint programs.

The relationship between hierarchy of authority and the number of joint programs is little affected by indicators of complexity, but somewhat more by the indicators of internal communication. (See Table 11–9.) On the other hand, the relationship between these two variables is reversed when the number of new programs is controlled, and the relationship is now in the expected direction, i.e., members of organizations with many joint programs having more control over individual work tasks. This finding buttresses our earlier interpretation that it was the dramatic increase of new programs that brought about less control over individual work decisions in organizations with many joint programs.

5. *There is no relationship between formalization and the number of joint programs.* Rules and regulations are important organizational mechanisms that are often used to insure the predictability of performance. There are several important aspects of rules as mechanisms of social control. One is the number of regulations specifying who is to do what, when, where, and why; this we call job codification.[8] A second is the diligency with which such rules are enforced; this we call rule observa-

TABLE 11-9
Partial Correlation Coefficients between Number of Joint Programs and Indicators of Centralization of Decision-making, Controlling for Indicators of Complexity, Innovation, and Internal Communication

Control variables	*Partial correlations between number of joint programs and participation in decision-making, controlling for the variable indicated*	*Partial correlations between number of joint programs and hierarchy of authority, controlling for the variable indicated*
Indicators of complexity		
Index of professional training	.27	.33
Index of professional activity	.01	.21
Number of occupations: 1967	−.10	.31
Indicator of organizational innovation		
Number of new programs: 1959-1966 (excluding new programs that are joint programs)	.20	−.28
Indicators of internal communication		
Number of committees	.16	.17
Number of committee meetings per month	.43	.22

tion.[9] A third is the degree to which the procedures defining a job are spelled out; this we call the index of specificity of jobs.[10]

Two of these three indicators of formalization, the degree of rule observation and the degree of specificity of jobs, had very small inverse relationships with the number of joint programs ($r = -0.06$ in each case), but each of these is hardly different from zero. The index of job codification was directly related to the number of joint programs ($r = 0.13$), but it too is little different from zero, although it is in the opposite direction to our expectation.

We conclude from these findings that formalization is unrelated to the degree of organizational interdependence, suggesting that either this kind of internal diversity is not very important or that we do not have valid measures of this phenomenon. However, there is some problem of interpretation because there was also some movement of the highly innovative organizations toward greater formalization. For example, there is a negative partial correlation between the number of joint programs and each of the indicators of formalization, i.e., job codification (partial $r = -0.11$), rule observation (partial $r = -0.37$), and degree of specificity of jobs (partial $r = -0.29$), when the number of new programs during the period 1959–1966 is partialled out.

CONTROLS FOR SIZE, AUSPICES, AGE, AND TECHNOLOGY

The sixteen organizations included in this study are, from one point of view, relatively homogeneous. All of them provide either psychiatric, social, or rehabilitation services of one kind or another. In comparison to economic organizations, they are indeed homogeneous. In addition, they are all located in a single metropolitan area. The reader might wonder, therefore, how far we can generalize from our study to other kinds of organizations or to organizations in other communities.

There are several ways in which some estimate of the generality can be made. One approach would be to divide the organizations into different categories, as was done in Tables 11–2 and 11–5. Here we emphasized the differences among a set of organizations that, considering the range of all organizations, are relatively homogeneous. The difficulty with this approach is that we are making comparisons among so few cases in each category.

An alternative approach is to look at some general variables that describe the conditions of all organization. The size of the organization is one such variable. Similarly the auspices of the organization, i.e., whether private or public, is another. And the age of the organization may also be an important factor here. Perrow (1967) has recently suggested another variable, the degree of routinization of technology. Undoubtedly there are others, but these represent some of the variables that one is likely to encounter in the literature and, therefore, are a good starting place for controls.

Since there were such great differences in the size of organizations in the study, a rank ordering of size is used. The correlation coefficient between size and the number of joint programs is positive and moderate ($r = 0.34$), which means that larger organizations have slightly more joint programs.

The auspices of the organization are measured by a dummy variable of private (1) versus public (0). The correlation coefficient between auspices and number of joint programs is 0.20, meaning that private organizations have slightly more joint programs.

The age of the organization was measured by constructing a trichotomous variable: (0), the organization was started in the post-Depression years (1938 to present); (1), the organization was started in the years following World War I (1918–1923); and (2), the organization was started prior to 1900. The correlation coefficient between age of the organization and the number of joint programs is —0.15, indicating that the younger organizations have slightly more joint programs.

Finally we looked at the type of technology, measured by the degree of routineness of work activities. By routineness of work we mean the degree to which organizational members have non-uniform work activities (Perrow, 1967; Woodward, 1965).[11] The correlation coefficient between routineness of work and the number of joint programs is —0.24, meaning that organizations with many joint programs have less routine technologies.

None of these four variables has strong relationships with the number of joint programs. When each of the relationships between the number of joint programs and the indicators of complexity, organizational innovation, internal communication, centralization, and formalization are controlled by each of these four variables separately, the relationships shown in Table 11–3 are little affected. (See Table 11–10.) This means that the factors of organizational size, auspices, age, and technology (as we have measured them) have little or no effect on the findings of this study.

TABLE 11-10
Partial Correlations between the Number of Joint Programs and Indicators of Complexity, Innovation, Internal Communication, Centralization, and Formalization, Controlling Separately for Organization Size, Auspices, Age, and Technology

	Partial correlation coefficient between number of joint programs and the organization characteristic indicated, controlling for			
	Size	*Auspices*	*Age*	*Technology*
Complexity				
Index of professional training	.35	.14	.16	.02
Index of professional activity	.56	.61	.64	.60
Number of occupations: 1967	.88	.86	.89	.86
Innovation				
Number of new programs: 1959-1966 (excluding new programs that are joint programs)	.73	.76	.74	.75
Internal communication				
Number of committees	.41	.45	.48	.48
Number of committee meetings per month	.81	.82	.82	.83
Centralization				
Index of participation in decision-making	.25	.27	.40	.18
Index of hierarchy of authority	.38	.35	.29	.33
Formalization				
Index of job codification	.18	.12	.07	.19
Index of rule observation	−.27	.00	−.10	−.02
Index of specificity of job	−.19	.03	−.16	.12

Discussions and Conclusions

We now return to the issues raised at the outset of this paper. How are organizational structure and interdependence related? How can the study of an organization and its environment be combined? What kinds of organizations are more cooperative and integrated with other organizations?

We noted that there is a greater degree of complexity, i.e., more occupational diversity and greater professionalism of staff, in those organizations with the most joint programs. The participation in joint programs is evidently one mechanism for adding new occupational specialties to the organization at a reduced cost. By combining the resources of the focal organization with one or more others, there is the possibility of adding new occupational specializations to the organizational roster. This is especially true because joint programs are likely to be of a highly specialized nature, providing services and activities that the focal organization cannot support alone.

The involvement of staff in interorganizational relationships introduces them to new ideas, new perspectives, and new techniques for solving organizational problems. The establishment of collegial relationships with comparable staff members of other organizations provides them with a comparative framework for understanding their own organizations. This is likely to affect their professional activities—attendance at meetings of professional societies—as well as reinforce professional standards of excellence. In these ways the involvement of organizations in joint programs has the effect of increasing the complexity of these social and health welfare organizations.

The heightened interdependence has other important implications for the internal structure of organizations. The partial or total commitment of organizational resources to other organizations is likely to affect various departments and the business office as well as the central programs of such an organization. Problems of coordination are likely to become particularly acute under such circumstances. The organization is forced to overcome these problems by heightening the frequency of internal communication. A more diverse committee structure and more committee meetings are mechanisms for handling such problems.

We would have expected that the heightened rates of communication would have resulted in more decentralization than appears to be the case. It is entirely possible that the problems of internal coordination may be reflected in some attempts to tighten the power structure, thus leading to less movement towards decentralization than we had expected. Also, the problems of internal coordination may be reflected in greater programming of the organization, or at least attempts in that direction, and this may be the reason why there is a small relationship between heightened interdependency, as we have measured it, and the degree of centralization.

Diversity in occupations (the degree of complexity) and power groups (the degree of decentralization) are related to the number of joint programs, but diversity in work, as reflected in the absence of rules, is not related to this measure of interdependence. In part this may be a consequence of the sudden increase in the rate of program innovation. But it may also be that the degree of formalization is not a good measure of diversity. It is the diversity of occupations, including their perspectives and self-interests, along with the representation of these points of view in a decentralized structure, that allows for diversity with the most critical consequences.

Our assumptions help to explain the steadily increasing frequency of organizational interdependency, especially that involving joint programs. As education levels increase, the division of labor proceeds (stimulated by research and technology), and organizations become more complex. As they do, they also become more innovative. The search for resources needed to support such innovations requires interdependent relations with other organizations. At first, these interdependencies may be established with organizations with different goals and in areas that are more tangential to the organization. Over time, however, it may be that cooperation among organizations will multiply, involving interdependencies in more critical areas, and involve organizations having more similar goals. It is scarcity of resources that forces organizations to enter into more cooperative activities with other organizations, thus creating greater integration of the organizations in a community structure. The long range consequence of this process will probably be a gradually heightened coordination in communities.

NOTES

1. For a discussion of some of the basic differences between individual and collective properties, see Lazarsfeld and Menzel (1960) and Coleman (1964).

2. One advantage of this procedure is that it allows for the cancellation of individual errors made by the job occupants of a particular position. It also allows for the elimination of certain idiosyncratic elements that result from the special privileges a particular occupant might have received as a consequence. An alternative procedure for computing organizational means is to weigh all respondents equally. These two procedures yield strikingly similar results for the variables reported in this paper. The product-moment correlation coefficients between the scores based on these two computational procedures were as follows for the variables indicated:

Hierarchy of authority	0.93
Participation in decision making	0.85
Job codification	0.89
Rule observation	0.89
Index of specificity of jobs	0.93
Index of routinization of technology	0.94
Professional training	0.90
Professional activity	0.93

3. It should be noted that our count of occupational specialties is not based on the number of specific job titles. Instead, each respondent was asked what he did and then this was coded according to the kind of professional activity and whether it was a specialty. This procedure was used for two reasons. First, it allows for comparability across organizations. Second, it avoids the problem of task specialization where one activity might be divided into many specific and separate tasks. (See Thompson, 1964.)

4. The index of professional activity, which ranged from 0 to 3 points, was computed as follows: (a) 1 point for belonging to a professional organization; (b) 1 point for attending at least two-thirds of the previous six meetings of any professional organization; (c) 1 point for the presentation of a paper or holding an office in any professional organization.

5. The index was scored as follows: (1) high school graduates or less education, with no professional training, received a score of 0; (2) high school graduates or less education, with some professional training, received a score of 1; (3) staff members with a college degree or some college, but an absence of other professional training, received a score of 2; (4) staff members with a college degree or some college, and the presence of some other professional training, received a score of 3; (5) the presence of training beyond a college degree, and the absence of other professional training, received a score of 4; (6) the presence of training beyond a college degree, and the presence of other professional training, received a score of 5.

6. The index of actual participation in decision making was based on the following four questions: (1) How frequently do you usually participate in the decision to hire new staff? (2) How frequently do you usually participate in the decisions on the promotion of any of the professional staff? (3) How frequently do you participate in decisions on the adoption of new policies? (4) How frequently do you participate in the decisions on the adoption of new programs? Respondents were assigned numerical scores from 1 (low participation) to 5 (high participation), depending on whether they answered "never," "sometimes," "often," or "always," respectively, to these questions. An average score on these questions was computed for each respondent, and then the data were aggregated into organizational scores as described above.

7. The empirical indicators of these concepts were derived from two scales developed by Richard Hall (1963), namely, hierarchy of authority and rules. The index of hierarchy of authority was computed by first averaging the replies of individual respondents to each of the following five statements: (1) There can be little action taken here until a supervisor approves a decision. (2) A person who wants to make his own decisions would be quickly discouraged here. (3) Even small matters have to be referred to someone higher up for a final answer. (4) I have to ask my boss before I do almost anything. (5) Any decision I make has to have my boss's approval. Responses could vary from 1 (definitely false) to 4 (definitely true). The individual scores were then combined into an organizational score as described above.

8. The index of job codification was based on responses to the following five statements: (1) A person can make his own decisions without checking with anybody else. (2) How things are done here is left up to the person doing the work. (3) People here are allowed to do almost as they please. (4) Most people here make their own rules on the job. Replies to these questions were scored from 1 (definitely true) to 4 (definitely false), and then each of the respondent's answers was averaged. Thus, a high score on this index means high job codification.

9. The index of rule observation was computed by averaging the responses to each of the following two statements: (1) The employees are constantly being checked on for rule violations. (2) People here feel as though they are constantly being watched, to see that they obey all the rules. Respondents' answers were coded from 1 (definitely false) to 4 (definitely true), and then the average score of each respondent on these items was computed. Organizational scores were computed as previously described. On this index, a high score means a high degree of rule observation.

10. The index of specificity of job was based on responses to the following six statements: (1) Whatever situation arises, we have procedures to follow in dealing with it. (2) Everyone has a specific job to do. (3) Going through the proper channels is constantly stressed. (4) The organization keeps a written record of everyone's job perform-

ance. (5) We are to follow strict operating procedures at all times. (6) Whenever we have a problem, we are supposed to go to the same person for an answer. Replies to these questions were scored from 1 (definitely false) to 4 (definitely true), and then the average score of each respondent on these items was computed as the other measures. A high score means a high degree of specificity of the job.

11. The index of routinization of technology was based on responses to the following five statements: (1) People here do the same job in the same way every day (reversed). (2) One thing people like around here is the variety of work. (3) Most jobs have something new happening every day. (4) There is something different to do every day. (5) Would you describe your job as being highly routine, somewhat routine, somewhat non-routine, or highly non-routine? The first four items were scored from 1 (definitely true) to 4 (definitely false). On the fifth item scores ranged from 1 (highly non-routine) to 4 (highly routine).

REFERENCES

Barth, Ernest A. T.
1963 "The causes and consequences of inter-agency conflict." Sociological Inquiry 33 (Winter): 51–57.

Black, Bertram J. and Harold M. Kase
1963 "Inter-agency cooperation in rehabilitation and mental health." Social Service Review 37 (March): 26–32.

Caplow, Theodore
1964 Principles of Organization. New York: Harcourt, Brace & World, Inc.

Clark, Burton R.
1965 "Interorganizational patterns in education." Administrative Science Quarterly 10 (September): 224–237.

Coleman, James S.
1964 "Research chronicle: the Adolescent society." Phillip Hammond (ed.), Sociologist at Work. New York: Basic Books.

Coser, Lewis
1956 The Functions of Social Conflict. Glencoe, Ill.: The Free Press of Glencoe.

Crozier, Michel
1964 The Bureaucratic Phenomenon. Chicago: The University of Chicago Press.

Dahrendorf, Ralf
1958 "Out of Utopia: toward a reorientation of sociological analysis." American Journal of Sociology 64 (September): 115–127.

Dill, William R.
1958 "Environment as an influence on managerial autonomy." Administrative Science Quarterly 2 (March): 409–443.
1962 "The impact of environment on organizational development." Pp. 94–109 in Sidney Mailick and Edward H. Van Ness (eds.), Concepts and Issues in Administrative Behavior. Englewood Cliffs, N.J.: Prentice-Hall, Inc.

Elling, R. H. and S. Halbsky
1961 "Organizational differentiation and support: a conceptual framework." Administrative Science Quarterly 6 (September): 185–209.

Emery, F. E. and E. L. Trist
1965 "The causal texture of organizational environment." Human Relations 18 (February): 21–31.

Evan, William M.
1966 "The organization-set: toward a theory of interorganizational relations." Pp. 173–191 in James D. Thompson (ed.), Approaches to Organizational Design. Pittsburgh, Pa.: University of Pittsburgh Press.
Form, William H. and Sigmund Nosow
1958 Community in Disaster. New York: Harper and Row.
Gouldner, Alvin
1959 "Reciprocity and autonomy in functional theory." Pp. 241–270 in Llewellyn Gross (ed.), Symposium on Sociological Theory. New York: Harper and Row.
Guetzkow, Harold
1950 "Interagency committee usage." Public Administration Review 10 (Summer): 190–196.
1966 "Relations among organizations." Pp. 13–44 in Raymond V. Bowers (ed.), Studies on Behavior in Organizations. Athens, Ga.: University of Georgia Press.
Hage, Jerald and Michael Aiken
1967 "Program change and organizational properties: a comparative analysis." American Journal of Sociology 72 (March): 503–519.
Hall, Richard
1963 "The concept of bureaucracy: an empirical assessment." American Journal of Sociology 69 (July): 32–40.
Harbison, Frederick H., E. Kochling, F. H. Cassell and H. C. Ruebman
1955 "Steel management on two continents." Management Science 2: 31–39.
Harrison, Paul M.
1959 Authority and Power in the Free Church Tradition. Princeton, N.J.: Princeton University Press.
Hawley, Amos H.
1951 Human Ecology, New York: The Ronald Press.
Homans, George
1950 The Human Group. New York: Harcourt, Brace and World, Inc.
Johns, Ray E. and David F. de Marche
1951 Community Organization and Agency Responsibility. New York: Association Press.
Kornhauser, William
1959 The Politics of Mass Society. Glencoe, Ill.: The Free Press of Glencoe.
Lawrence, Paul R. and Jay W. Lorsch
1967 Organization and Environment. Boston: Graduate School of Business Administration, Harvard University.
Lazarsfeld, Paul and Herbert Menzel
1960 "On the relation between individual and collective properties." Pp. 422–440 in Amitai Etzioni (ed.), Complex Organizations: A Sociological Reader. New York: The Macmillan Company.
Lefton, Mark and William Rosengren
1966 "Organizations and clients: lateral and longitudinal dimensions." American Sociological Review 31 (December): 802–810.
Levine, Sol and Paul E. White
1961 "Exchange as a conceptual framework for the study of interorganizational relationships." Administrative Science Quarterly 5 (March): 583–601.
1963 "The community of health organizations." Pp. 321–347 in Howard E. Freeman, S. E. Levine, and Leo G. Reeder (eds.), Handbook of Medical Sociology. Englewood Cliffs, N.J.: Prentice-Hall.

Levine, Sol, Paul E. White and Benjamin D. Paul
1963 "Community interorganizational problems in providing medical care and social services." American Journal of Public Health 53 (August): 1183–1195.
Litwak, Eugene
1961 "Models of bureaucracy which permit conflict." American Journal of Sociology 67 (September): 177–184.
Litwak, Eugene and Lydia F. Hylton
1962 "Interorganizational analysis: A hypothesis on coordinating agencies." Administrative Science Quarterly 6 (March): 395–426.
Miller, Walter B.
1958 "Inter-institutional conflict as a major impediment to delinquency prevention." Human Organization 17 (Fall): 20–23.
Morris, Robert and Ollie A. Randall
1965 "Planning and organization of community services for the elderly." Social Work 10 (January): 96–102.
Perrow, Charles
1967 "A framework for the comparative analysis of organizations." American Sociological Review 32 (April): 194–203.
Reid, William
1964 "Interagency coordination in delinquency prevention and control." Social Service Review 38 (December 1964): 418–428.
Richardson, Stephen A.
1959 "Organizational contrasts on British and American ships." Administrative Science Quarterly 1 (September): 189–207.
Ridgeway, V. F.
1957 "Administration of manufacturer-dealer systems." Administrative Science Quarterly 1 (June): 464–483.
Rushing, William A.
1967 "The effects of industry size and division of labor on administration." Administrative Science Quarterly 12 (September): 273–295.
Selznick, Philip
1949 TVA and the Grass Roots. Berkeley, Cal.: University of California Press.
Simpson, Richard L. and William H. Gulley
1962 "Goals, environmental pressures, and organizational characteristics." American Sociological Review 27 (June): 344–351.
Terreberry, Shirley
1968 "The evolution of organizational environments." Administrative Science Review 12 (March): 590–613.
Thompson, James D.
1962 "Organizations and output transactions." American Journal of Sociology 68 (November): 309–324.
1966 Organizations in Action. New York: McGraw-Hill.
Thompson, James D. and William J. McEwen
1958 "Organizational goals and environment: goal-setting as an interaction process." American Sociological Review 23 (February): 23–31.
Thompson, Victor R.
1961 Modern Organizations. New York: Alfred A. Knopf, Inc.
Tocqueville, Alexis de
1945 Democracy in America. New York: Alfred A. Knopf, Inc.
Warren, Roland L.
1965 "The impact of new designs of community organization." Child Welfare 44 (November): 494–500.

1967 "The interorganizational field as a focus for investigation." Administrative Science Quarterly 12 (December): 396–419.
Wilensky, Harold L.
1967 Organizational Intelligence. New York: Basic Books, Inc.
Willer, David
1967 Scientific Sociology: Theory and Method. Englewood Cliffs, New Jersey: Prentice-Hall.
Wilson, James Q.
1966 "Innovation in organization: notes toward a theory." Pp. 193–218 in Approaches to Organizational Design. Pittsbugh, Pa.: University of Pittsburgh Press.
Woodward, Joan
1965 Industrial Organization. London: Oxford University Press.
Wrong, Dennis
1961 "The oversocialized conception of man in modern society." American Sociological Review 26 (April): 183–193.
Yuchtman, Ephraim and Stanley E. Seashore
1967 "A system resource approach to organizational effectiveness." American Sociological Review 32 (December): 891–903.

12

JOSEPH W. SCOTT AND
MOHAMED EL-ASSAL

MULTIVERSITY, UNIVERSITY SIZE, UNIVERSITY QUALITY, AND STUDENT PROTEST: AN EMPIRICAL STUDY

Introduction

Berkeley and Columbia (Draper, 1965; The Graduate Student Society, 1968) have become symbols of a militant generation of college students and of student activism. Student activism, characteristic of the 1960's has become a defined "social problem." Since the Berkeley demonstrations of 1964, such activism has increased in number of incidents and in the degree of militancy. Recently students have not only employed the "sit-in," but also have captured buildings and held administrators as hostages, sometimes until they agreed to demands of the students or until the police arrived.

There is ample evidence that current student protest movement in the U.S. had its origin in the 1950's (Lipset, 1966). Widespread anti-ROTC demonstrations occurred largely in the late '50's, indicating that the students were not as "silent" and as contented as some writers would have us believe (Scott, 1968). Yet, notwithstanding such signs of the future, no forecast was made of the activism and militancy of students present today on U.S. campuses.

Very few systematic studies on the student movement have been re-

Reprinted from Joseph W. Scott and Mohamed El-Assal, "Multiversity, University Size, University Quality, and Student Protest: An Empirical Study," *American Sociological Review* 34–5 (October 1969), pp. 702–709.

ported (Lyonns, 1965; Somers, 1965; Watts and Whittaker, 1966; Flacks, 1967; Trent and Craise, 1967; Westby and Braungart, 1966; Solomon and Fishman, 1964; Peterson, 1966; Sasajima *et al.*, 1967; Schiff, 1964). Even after Berkeley, few, if any, studies on student protest have appeared in the major sociological journals.[1] Many journalistic accounts and commentaries on student activism were written, but these have not been based on methodical and empirical research. In the absence of such research, the nature of the protest movement is a matter largely for speculations and opinions.

Although the general subject of student demonstrations stands in need of research, some aspects are more deserving of concentration than others. To date, most authors have dealt with the personal characteristics of the activists themselves. These authors have compared student activists with other college groups, with national samples of college students, and with non-college samples. In some of these studies, the student activists have been given personality inventories and have been questioned on their private values and political ideologies. In addition, some parents and grandparents have been interviewed. Yet, these studies still leave a significant gap in our knowledge. University structures themselves have been studied, but few reports have been discovered by the writers; one report found is by Peterson (1966). The apparent scarcity of published material is somewhat inconsistent with the importance of university administrative structures as a factor in student grievances. Specifically, the increasing social heterogeneity of the student bodies and bureaucratization of the administrative structures are of crucial importance because they are two principal conditions stimulating student unrest and moving student activists to protest demonstrations. For these reasons the university structures in relation to their environments should be intensively studied. What has been reported about the relationship of structures to their environments has been written largely by behavioral scientists (Lipset and Altbach, 1966; Katz and Sanford, 1965; Reisman, 1958; Keniston, 1962; Springer, 1968; Seabury, 1966; Brown, 1960; Fishman and Solomon, 1963).

Since so many speculations have been advanced, we believe that systematic research is required to determine which speculations are best supported by empirical data. For example, permissive family socialization (Flacks, 1967; Heist, 1965; Westby, 1966), selective university recruitment and socialization (Kaplan, 1965; Lipset, 1966; Lipset and Seabury, 1965), university size and heterogeneity (Peterson, 1966 and 1968), university opportunity structures for protest (Starobin, 1965; Peterson, 1968), and university bureaucratization (Nelken, 1965; Keniston, 1967) have all been subjects of speculation as causes of student unrest and rebellion. Our study tries to sort out the strength of some of these variables.

The Problem

Since 1960, most American universities have changed structurally and socially, and some of these changes have given rise to conflicts between students and administrators. During the time that many of these schools have grown structurally from simple teachers colleges to complex universities, their number of mature graduate and undergraduate students have greatly increased. Such students do not want their campus activities and classwork dictated, regulated, managed or rigidly preplanned. But paradoxically, as a larger percentage of college and university students are married, mobile, financially independent, and more insistent on personal freedoms, many administrations have become more bureaucratic, and regimentation of the educational process has increased. Thus, many administrators have unwittingly been drawn into conflict with the activist students. The issues and rhetoric of student protest demonstrations as reported in the mass media lend support to this conclusion. The nature of this conflict has been noted by Clark Kerr (1965), former president of the University of California, who spoke of an incipient revolt against the multiversity as follows:

If the alumni are concerned, the undergraduate students are restless. Recent changes in the American university have done them little good—lower teaching loads for the faculty, larger classes, the use of substitute teachers for the regular faculty, the choice of faculty members based on research accomplishments rather than instructional capacity, the fragmentation of knowledge into endless subdivisions. There is an incipient revolt of undergraduate students against the faculty; the revolt that used to be against the faculty in *loco parentis* is now against the faculty *in absentia*. The students find themselves under a blanket of impersonal rules for admissions, for scholarships, for examinations, for degrees. It is interesting to watch how a faculty intent on few rules for itself can fashion such a plethora of them for the students. The students also want to be treated as distinct individuals.

Continuing in the same vein, Kerr held that the typical state university had become large, complex, and diversified; teaching had become less important than research; faculty members had become less members of the particular university and more colleagues within their academic disciplines; teaching had become minimal for the most highly paid faculty members; classes had become generally larger; the students themselves had become so vast in number and so heterogeneous that the campus for them often seemed confusing, full of dilemmas, fraught with problems of belonging and security.

Kerr in the statement above offers us a research proposition which may be tested. First, let us postulate that the more complex the formal structure is, the more likely is the administration to be bureaucratic as opposed to primary and patrimonial. Accordingly, the more bureaucratic the educational institution, the more structurally separated are the students from administrators, faculty, and students; and the more the students are personally separated from administrators, faculty members, and other students by structural complexity and social heterogeneity, the more likely the students will feel separated, neglected, manipulated, and dehumanized to the extent that they will engage in protest activities. Given these premises, we hypothesize that the more nearly a university constitutes a "multiversity," the higher the rate of protest demonstration. To test the stated research hypothesis, we had to correlate the degrees of formal complexity and social heterogeneity with the number of student demonstrations, introducing at various points the intervening variables of institutional size, institutional quality, and size of community in which the educational institution is located.

Method

First, we selected a *purposive sample* of 104 state-supported non-technical and non-specialized colleges and universities that reflect the diverse (1) sizes of such institutions, (2) sizes of resident communities, and (3) geographic parts of the United States. Secondly, we compiled separate socio-demographic profiles for each school (college or university) using secondary sources, namely, *The College Blue Book: American Universities and Colleges*, and *The Digest of Educational Statistics*, and other documents published by the U.S. Department of Health, Education, and Welfare. Thirdly, we gathered a list of the number and types of demonstrations occurring at each school from January 1, 1964 to December 31, 1965. Each dean of students and each editor of the school newspaper was then sent a questionnaire containing the following categories of information to which they were to reply: (1) the issues over which any demonstrations occurred, (2) the duration of each demonstration, (3) actions of demonstrators, i.e., violent or peaceful, (4) organizations involved, (5) actions taken by the administration, (6) the total number of demonstrations.

These questionnaires were mailed to the representatives of the schools in our sample. The receipt of one or more questionnaires from the same school was considered as a single report for the school. When one dean and one student editor were diametrically opposed in their reporting, we wrote to them for clarification. The mail-back rate was as follows:

Schools not responding	*Schools reporting no demonstrations*	*Schools reporting some demonstrations*	*Total*
35	25	44	104

Compared with other studies requiring mail-back questionnaires, the ratio of questionnaires returned was quite favorable. The mail-back rates were nearly equal when schools with an enrollment over ten thousand were compared to those under ten thousand. The questionnaire responses came from schools in 42 states from all geographical sections of the United States, namely, New England, Middle Atlantic, North Central, Southwest, Rocky Mountain, and Far West.

The *independent variable* in this study was "multiversity"—an index combining structural complexity and social heterogeneity. The *dependent variable* was the number of demonstrations. The *intervening variables* were the size and quality of the institution, and the size of the community in which the institution was located.

"Multiversity" was operationalized as composite index based on these factors: (1) the number of departments granting doctoral degrees, (2) the number of departments granting masters degrees, (3) the number of departments granting bachelors degrees, (4) the number of departments offering first professional degrees, (5) the ratio of dormitory to non-dormitory students, (6) the ratio of outstate to instate students, (7) the ratio of foreign students to native born students, (8) the ratio of students to professors, and (9) the ratio of undergraduates to graduate students. The first four variables were clearly indicators of administrative complexity: span of control, horizontal and vertical specialization, the number of hierarchies of authority, the degrees of segmentation and coordination, and the number of rules, regulations, and lines of communication. The second five variables were indicators of administrative complexity and social heterogeneity of the student body. These variables not only indicate the diverse human elements which must be administratively allocated, integrated, and managed daily in order for the university to carry on, but they also indicate socially the extent of secondariness in the daily interpersonal contacts among the students themselves.

The multiversity index was derived by first recording the scores and ratios for each school using these nine dimensions. Secondly, we arrayed scores and ratios from high to low. Thirdly, we reduced each array to dichotomized categories using the medians as cutting points. Fourthly, we ran a Guttman scale analysis using these dichotomous scores for the nine variables. The result was a quasi-scale with a coefficient of reproducibility of .84 and a minimal marginal coefficient of reproducibility of .51. No items were eliminated in order to raise the coefficient of reproducibility, since we were interested only in reliably ranking schools from complex to simple by our composite index. The quasi-scale scores varied from ten to one. The arrayed scores resulted in a bimodal distribution with about 30% of the cases receiving a score of one and about 38% receiving a score of ten. The other scores were evenly distributed between these extremes. Finally, using the median of this array of scores, we divided the schools into two categories: "complex" versus "simple" schools. This dichotomy will be our short way of designating "multiversities" versus "non-multiversities" respectively.

Abstractly a "complex" school would be above the median in numbers of departments granting doctoral, masters, bachelors and professional degrees, and in numbers of non-dormitory, outstate, foreign, and graduate students, as well as in numbers of professors to students.

Definitions

Size of school in this study was the number of full-time equivalent students. The schools were categorized as follows: 15,000 or more; 10,000–15,000; 7,000–10,000; 4,000–7,000; and 2,000–4,000. *Quality* of school was computed by summing the schools' positions in three arrays: the ratio of the number of library books to all students, the percent of full-time faculty with doctoral degrees, and the percentage of graduates receiving national scholarships and honors. When a school was above the median, it received a score of one; if it was below, it received a score of two. The quality index varied from three (high quality) to six (low quality).

Finally, *size of community* was defined as the total population of the community according to the official census of 1960. We began with four categories: 500,000 or more; 50,000–500,000; 15,000–50,000; and 15,000 or less.

With the schools categorized as "simple" and "complex," we analyzed the relationship among the variables: complexity, size, and quality of schools, size of community in which the schools were located, and number of student demonstrations.

TABLE 12-1
School Characteristics and Schools Reporting Student Demonstrations

School characteristics	*Intervening variables*	*Schools reporting demon-strations*	*Schools reporting no demon-strations*	*N*	
Complex institutions		87%	13%	32	G = .80
Simple institutions		43	57	37	
School size					
Large (10,000+)		96	4	26	G = .94
Small (10,000–)		44	56	43	
Community size					
Large city (50,000+)		74	26	27	G = .36
Small city (50,000–)		57	43	42	
High quality institutions		85	15	33	G = .69
Low quality institutions		44	56	36	
School size					
Mostly complex	10,000+	96	4	23	
	10,000–	67	33	9	G = .82
Mostly simple	10,000+	100	0	3	
	10,000–	38	62	34	
Community size					
Mostly complex	50,000+	94	6	16	
	50,000–	81	19	16	G = .66
Mostly simple	50,000+	42	58	26	
	50,000–	45	55	11	
School quality					
Mostly complex	High quality	89	11	27	
	Low quality	80	20	5	G = .75
Mostly simple	High quality	67	33	6	
	Low quality	39	61	31	

Findings

COMPLEXITY

The great majority of the complex schools reportedly had some demonstrations, while the great majority of the simple schools reportedly did not have demonstrations. The gamma coefficient representing this association was .08. Complex schools had 2.44 demonstrations per school reporting, and simple schools had .75 demonstrations per school reporting. (See Table 12–1.)

SIZE OF SCHOOL

Generally, the schools with ten thousand or more students had the most demonstrations. Over 90% reported having had some demonstrations. The medium-sized and small schools usually reported no demonstrations. The gamma coefficient for this association was .94. The large schools had 2.89 demonstrations per school, and the small schools had .72 demonstrations per school.

SIZE OF SCHOOL AND COMPLEXITY

The schools with enrollments of 10,000 or more students fell disproportionately in the complex category, and the schools of fewer than 10,000 students fell disproportionately in the simple category. However, *regardless of complexity, all the large schools, except one, reported having had some demonstrations.* But, among the smaller schools, complexity exerted an appreciable influence: the great majority of small schools in the complex category had demonstrations, whereas the great majority of the small schools in the simple category did not have demonstrations. The gamma coefficient representing this association was .82.

SIZE OF COMMUNITY

All in all, the schools in large- and medium-sized communities had disproportionately more demonstrations. The great majority of schools in small-sized towns did not have demonstrations. The gamma coefficient representing this association was .36.

SIZE OF COMMUNITY AND COMPLEXITY

The great majority of the complex schools in both large and small communities had demonstrations. However, a substantial majority of simple schools had not had demonstrations, especially when they were located in communities of fewer than 50,000 persons. The gamma coefficient representing this association was .66.

QUALITY OF SCHOOL

By and large, the schools of highest quality (as defined) had disproportionately more demonstrations than schools of the lowest quality. A great majority of high quality schools had demonstrations, while a majority of low quality schools had no demonstrations. The gamma coefficient representing this association was .72. There were 2.36 demonstrations per high quality school and .75 demonstration per low quality school.

QUALITY OF SCHOOL AND COMPLEXITY

The great majority of complex schools regardless of quality had demonstrations. By contrast, among the simple schools, two-thirds of the high quality schools had demonstrations; very few simple, low quality schools had demonstrations. In sum, high quality and complexity increased greatly the probability of schools having demonstrations but low quality and simplicity reduced greatly the probability of schools having them. The gamma coefficient representing this association was .75.

Introducing all the intervening variables at the same time did not give us enough cases in some cells to indicate all interactive effects. This multivariate table, however, did indicate that 100% of the complex, large, high quality schools had demonstrations, while only 34% of the simple, small, low quality schools had them. Half of these small schools had only one demonstration each during the period surveyed.

Our hypothesis that *as an institution becomes more like a multiversity the rate of demonstration also becomes higher appears to have been supported*. School size seems to give rise to those structural and social features which lead us to term an institution with such characteristics "a multiveristy." School size was, therefore, found to be the most consistent predictor of student demonstrations.

In terms of predictive power, our non-parametric analysis revealed that school size was first; complexity was second; quality was third; and size of community was fourth. To get some indication of their relative numerical weights as predictive variables, we completed another analysis of our data.[2]

In Table 12–2 we have presented the results of a stepwise regression analysis. The table contains the multiple correlation coefficients, the beta coefficients, and simple zero-order inter-correlations coefficients. From this

TABLE 12-2
Stepwise Multiple Correlation Coefficients, Beta Coefficients and Zero-order Coefficients for School Size, Complexity, School Quality, Community Size and Number of Demonstrations

Independent variables	*Number of demonstrations*	*% Variance explained*
	R	
School size	.580	33.6
School size and complexity	.591	34.9
School size, complexity, and quality	.593	35.2
School size, complexity, quality, and community size	.594	35.3

Independent variables	*Beta coefficients*
School size	.430
Complexity	.119
Quality	.080
Community size	.038

	Matrix of zero-order correlation coefficients				
Variables	*(1)*	*(2)*	*(3)*	*(4)*	*(5)*
(1) Complexity	—	73	72	.27	.60
(2) Quality	—	—	.58	.17	.42
(3) School size	—	—	—	.46	.58
(4) Community size	—	—	—	—	.28
(5) Number of demonstrations	—	—	—	—	—

analysis, the ordinal relationships among the independent variables are clarified. School size accounted for almost all the explained variance. Complexity was heavily dependent on school size and thus accounted for only a small part of the total variance. School quality and community size accounted for almost none of the variance. Although these data do not meet all the assumptions of this regression analysis, we nevertheless believe the results to be suggestive. In the next section we present a paradigm to explain these findings.

INCREASING SIZE AND BUREAUCRATIZATION OF UNIVERSITIES

In becoming large, high quality, and heterogeneous, many schools had to expand and to formalize their administrative structures in order to coordinate large diverse numbers of communication activities and people in the massive educational enterprises of teaching, research, and public service. Accordingly, they expanded their administrative staff personnel such as vice-presidents, deans, associate deans, department chairmen, associate department chairmen, administrative assistants, secretaries, executive secretaries, clerks, clerical helpers, and consultants. Simultaneously, they formalized and routinized their administrative procedures in order to coordinate and to regulate the granting of examinations, of degrees, of stipends, of scholarships, research on human subjects, teaching and research facilities, housing, extracurricular organizations, teaching schedules, speakers, sports activities, and hiring, termination, promotion and evaluation of personnel.

This routinization by formalization "caused" students, faculty, and staff alike to be treated more and more as impersonal and processable categories of people characterized by some "common factor." When such personnel (the students in particular) had grievances, these grievances were routinely "processed" by administrative assistants, IBM cards, letters, or standardized forms and the like, thereby effectively closing off these people from the top of the university hierarchy. For students to receive some direct *personal* contact from higher administrators when they had grievances for which they sought redress, they sometimes had to engage in illegal or extralegal protest actions, e.g., collective demonstrations and disorders, which were not easily processed by daily routines, not easily mediated by impersonal mediators or communication, not easily ignored, and finally not easily delegated to lower-level bureaucrats by the top-level university administrators. Large, complex, high quality schools, being more likely to use autocratic and bureaucratic means to redress student grievances and being more likely to provide little or no opportunity for the concerned students to directly vote on the outcome of their grievances, encouraged and increased the likelihood of student protest demonstrations (Lipset, 1966; Nelken, 1965). The probability of demonstrations was also increased because at the same time that these particular schools heightened the experience of administrative autocracy, administrative regimentation, and social alienation, they also attracted, recruited and socialized more radicalized and politicalized students who were inclined to change these very structural conditions wherever they found them in society (Kaplan, 1965; Keniston, 1967; Peterson, 1966; Savio, 1965).

Conclusions

Large, complex, high quality schools had a much higher rate of demonstrations per school than small, simple, low quality schools. These large schools did not, however, produce more demonstrations per 1,000 students than the small schools, indicating that demonstrations are not a result of number alone. The interaction effect between school size, administrative complexity and social heterogeneity was found to be fundamental to understanding some social conditions associated with student unrest and student protest in the U.S. today.

NOTES

1. The authors could find no studies on the student revolution in the *American Sociological Review, American Journal of Sociology, Social Problems, Sociometry, Social Forces, Social Science* and *Social Research* and others, during the period of 1950–1967.

2. One of the ASR referees raised the question of whether or not larger schools (10,000+) really have a higher rate of demonstrations per 1,000 students than what would be normally expected. In calculating the rate of demonstrations per 1,000 students for each school, we learned that more large schools had demonstrations than the small schools, but they were on the whole not higher rates per 1,000 students. Large size does not produce a higher rate of demonstrations than small size; large size is simply more likely to be associated with the occurrence of student protest demonstrations. Complexity is positively associated with student protest demonstrations among both large schools (10,000+) and small schools (10,000−). The relationship is more definitive among small schools than among large schools. Of the 26 large schools reporting, 23 were clearly complex, and 3 were clearly simple. Since all large schools but one had demonstrations, complexity does not stand out as vividly among large schools as it does among small schools where there is a sharp increase in demonstrations as small schools become more complex.

REFERENCES

Brown, Donald R.
1960 "Social changes and the college student: A symposium." The Educational Record (October): 329–358.

Draper, Hal
1965 Berkeley: The New Student Revolt. New York: Grove.

Fishman, Jacob R., and Fredric Solomon
1963 "Youth and social action: Perspectives on the student sit-in movement." American Journal of Orthopsychiatry 33 (May): 872–82.

Flacks, Richard
1967 "The liberated generation: Explanation of the roots of student protest." The Journal of Social Issues 23 (July): 52–75.

Heist, Paul
1965 "Intellect and commitment: The faces of discontent." Berkeley: Center for the Study of Higher Education.

Kaplan, Samuel
1965 "The revolt of the elite: Sources of the FSM victory." The Graduate Student Journal 4 (Spring): 27–29.

Katz, Joseph and Nevitt Sanford
1965 "Causes of the student revolution." Saturday Review (December): 64–79.

Keniston, Kenneth
1962 "American students and the political revival." The American Scholar 32 (Winter): 40–64.
1967 "The sources of student dissent." The Journal of Social Issues (July): 108–137.

Kerr, Clark
1965 "Selections from The Uses of the University." Pp. 38–60 in S. M. Lipset and S. S. Wolin (eds.), The Berkeley Student Revolt. New York: Doubleday.

Lipset, Seymour M.
1966 "Student opposition in the United States." Government and Opposition (April): 351–374.

Lipset, Seymour M. and Philip G. Altbach
1966 "Student politics and higher education in the United States." Comparative Education Review (June): 321–49.

Lipset, Seymour M. and Paul Seabury
1965 "The lesson of Berkeley." The Reporter (January): 36–40.

Lipset, Seymour M. and Sheldon S. Wolin
1965 The Berkeley Student Revolt. New York: Doubleday.

Lyonns, Glen
1965 "The police car demonstration: A survey of participants." Pp. 519–530 in S. M. Lipset and S. S. Wolin (eds.), The Berkeley Student Revolt. New York: Doubleday.

Nelken, Michael
1965 "My mind is not property." The Graduate Student Journal (Spring): 30–34.

Peterson, Richard E.
1966 The Scope of Organized Student Protest. Princeton: Education Testing Service.
1968 "The student left in American higher education." Daedalus (Winter): 293–317.

Reisman, David
1958 "The college student in an age of organization." Chicago Review 12 (Autumn): 50–58.

Sasajima, Masa, Jonius A. Davis, and Richard E. Peterson
1967 "Organized student protest and institutional climate." Princeton: Educational Testing Service.

Savio, Mario
1965 "An end to history." Pp. 216–219 in S. M. Lipset and S. S. Wolin (eds), The Berkeley Student Revolt. New York: Doubleday.

Schiff, Lawrence F.
1964 "The obedient rebels: A Study of college conversions to conservatism." The Journal of Social Issues 20 (October): 75–94.

Scott, Joseph W.
1968 "ROTC recruitment, student protests, and change in the military establishment." University of Toledo: Mimeographed.

Seabury, Paul
1966 "Berlin and Berkeley." Comparative Education Review (June): 350–58.

Solomon, Frederic and Jacob R. Fishman
1964 "Youth and peace: A psychological study of peace demonstrators in Washington, D. C." The Journal of Social Issues 20 (October): 54–73.

Somers, Robert H.
1965 "The mainsprings of the rebellion: A survey of Berkeley students in November 1964." Pp. 530–557 in S. M. Lipset and S. S. Wolin (eds.), The Berkeley Student Revolt. New York: Doubleday.
Springer, George P.
1968 "Universities in flux." Comparative Education Review (February): 28–38.
Starobin, Robert
1965 "Graduate students and the FSM." The Graduate Student Journal (Spring): 25.
The Graduate Student Society
1968 "The Columbia student strike." New York: Columbia University
Trent, James W. and Judith L. Craise
1967 "Commitment and conformity in the American College." The Journal of Social Issues 23 (July): 34–51.
Watts, William and Dave Whittaker
1966 "Free speech advocates at Berkeley." The Journal of Applied Behavioral Science 2 (Winter): 41–62.
Westby, David L., and Richard G. Braungart
1966 "Class and politics in the family backgrounds of student political activists." American Sociological Review 31 (April): 690–92.

Index